Mike Law has spent his career developing the most advanced racing machines the world has ever seen. In this environment, every millisecond counts, and tiny innovations can lead to a crucial advantage against the fierce competition. Understanding where to focus your time and energy is the difference between the podium and the back of the grid.

In ACE Thinking, Mike describes the universality between the different tasks we participate in every day, breaking down complex ideas from science and mathematics into basic concepts that can be used in task planning, execution and optimisation. These ideas build into the ACE Model; a framework capable of improving the outcome of anything we set our minds to.

About the Author

Mike attended Loughborough University, where he studied Automotive Engineering, finishing as top of his graduating class and winning the Henry Ford prize. After completing his studies, Mike went straight into a career in motorsport, working in the Vehicle Dynamics departments for two of the most successful teams in the history of Formula 1.

Outside his career, Mike has worked with the Smallpeice Trust, helping to select candidates for the Arkwright Scholarship, an award that he won during his time at school.

Mike currently lives in Hampshire with his wife Hannah and their children, Jack and Rowan.

ACE THINKING

Life Lessons from Engineering the Ultimate Racing Cars

Mike Law

Derry Down

First edition September 2022 by Derry Down

This is a work of creative nonfiction. Some parts have been fictionalised in varying degrees for the purpose of clarity.

Copyright © Mike Law, 2022

All rights reserved. No part of this book may be reproduced or used in any manner without written permission of the copyright owner except for the use of quotations in a book review.

Edited by Hannah Joyce
Cover design by John Law
Pictures from Shutterstock

ISBN
979-8-4252-5852-6 (hardback)
979-8-5188-3892-5 (paperback)

The author bears no responsibility for the persistence or accuracy of the URLs for external or third-party internet websites referred to in this book and does not guarantee that any content on such websites is, or will remain, accurate or appropriate.

"It would be a waste of life to do nothing with one's ability, for I feel life is measured in achievement, not in years alone."

Bruce McLaren

Foreword

Formula 1, like most other industries, has been through an incredible revolution over the last 20 years or so that I have been working within it, driven by the information revolution. The complexity of the cars that you see on track is matched by the complexity of engineering that goes into them in the factory - employing cutting edge aerodynamic, vehicle dynamics and design tools with an obsessive attention to detail. There is so much data and information available that it is easy to become lost or misdirected in a project and, despite best intentions, miss the primary purpose of all Formula 1 engineering – namely to make the car and driver faster and thus beat your competition.

In my time as an F1 Race Engineer, I relied upon high quality efficient information to make decisions, whether about setup, strategy or run planning, and most of this information came from support engineers working back at the factory. It was in this capacity that I first worked with Mike and was quickly impressed by his refreshing approach to not just collaboratively solving the problems we had, but also improving our processes to find performance more efficiently in the future.

Since moving into a factory-based engineering role myself, I work with Mike more closely and am able to appreciate more of his methods. I was therefore very keen to read this book, ACE Thinking, and see what was behind his approach. I was not disappointed: Mike pulls upon a great number of theories and techniques associated with engineering, human behaviour and control theory and brings them together.

There is much fantastic research and reading material out there, but like most of us, I do not have time to read it all. This book brings key aspects together into a coherent process that can be applied to any problem or project, be that engineering or not, whilst also pointing the reader towards further information should they wish to go into more detail on a specific topic. (As well as being a great resource in its own right, it is also a quality reading list.)

I was therefore honoured to be asked by Mike to write this foreword and I hope that by explaining how he approaches problems that you, the reader, will be able to take some of this approach to your own profession (or indeed social or family life!).

Mark Temple
Senior F1 Car Performance Engineer and former F1 Race Engineer

Introduction

1

Cold Dinner and Concert Pianists

The end of the working day was fast approaching when I found myself summoned to yet another meeting. This one had been called at short notice to discuss an area of car performance which my group was responsible for. Unfortunately, things hadn't gone quite as well as we'd hoped during the last race; a fact made quite plain by the driver's numerous and heart-felt complaints about the car's handling.

This can happen from time to time, and it brings any hopes of an easy few days crashing down around you. In this case, the simulation tools we'd been using before the race hadn't identified the problem we'd ended up with. The tools will never be perfect, and we would normally have contingencies for this kind of thing, but throw logistical problems into the mix and it was obvious that we were going to struggle.

This type of meeting can be a very intimidating affair, filled with the sort of people who you'd been watching on TV twenty years earlier when you were dreaming of a career in Formula 1. It was addressing this type of situation that had got them to their current positions, and you'd need to be pretty switched on if you wanted to make any kind of impression. Despite this, and the challenges of the race, I felt reasonably confident that my group understood where we'd gone wrong and, more importantly, how to put steps in place to make sure it didn't happen again. Indeed, I bounced into the meeting confident I had a comprehensive plan that would convince the attendees that everything was perfectly under control.

After we'd filed our way into the meeting room, there were brief exchanges of pleasantries before we got down to the matter at hand. As is the way with these things, the most senior people in the room tend to dictate what's discussed, and it's not hard to understand why. They've seen a whole race weekend blighted by a persistent issue that distracted from other areas of performance, and they are under pressure from their bosses to fix it quickly. It was clear that any repeat of this would not be tolerated.

We started off reviewing the events from the weekend, summarising the causes and consequences of the problems we'd had, before moving onto the critical matter – what we were going to do about it. As always with this level of attendee, there were no shortage of reasonable suggestions for approaches we could take. These ranged from those based in the team's prior experience of tackling this kind of problem, through to more adventurous ones that relied on cutting-edge new techniques from our own research and development activities. Any one of these suggestions was likely to have led us to a better approach for the future, and this might have been the end of it, except that I knew that we could do better…

In the years and months leading up to this point in time, I'd been developing my own approach to the problems we encounter in this sport. This approach was one that I'd used to stand out from the other candidates for my position, largely because I wasn't able to compete with them through my knowledge of vehicle dynamics and performance alone! It was this that meant I viewed problems slightly differently from others around the table, and I chose this as the opportunity to make a new proposal that was taken from this alternative way of thinking. Alas, despite my best efforts to explain it concisely, avoiding use of complex terminology that wouldn't have been familiar to the assembled audience, the response I was met with could probably be best described as bewilderment. Clearly, I had asked a little too much of everyone to consider something so alien. This had the effect of delaying the end of the meeting whilst the key differences between what I was proposing and the familiar reasoning of everyone else in the room was unpacked. Because of this, it was getting very late before everything was wrapped up. Another day for apologies to my wife and children upon finally arriving home!

This meeting represented something of a turning point for me, and not only because I clearly needed to work on my communication skills. After the dust had settled, I felt frustrated that I hadn't succeeded in directing our solution away from established methods, towards my preferred way of working. My approach was something I had built up over time through an enthusiastic pursuit of knowledge over a variety of mathematical and scientific fields, including an in-depth study into the scientific process itself. It was a system that promoted humility in the face of the unbelievable challenges in engineering and drew strict boundaries on what constituted true *learning*. The techniques included were something I had strived to advocate to the team of engineers that I am responsible for and I'm confident that they were a key factor in their ability to wield intimidatingly technical concepts with confidence, whilst delivering results in good time and with minimal resource.

At the same time though, I knew I lacked the ability to describe the way we worked under a single term that adequately captured our philosophy. This term, had it existed, could have been my secret weapon during the meeting; the empowerment I needed to field almost any question thrown at me. This kind of term seemed possible in other areas of life, for example in politics you can, should you choose, label yourself as either conservative or liberal. In wider society you can describe yourself as an environmentalist, a feminist, or anything else which morally drives you. These terms suggest a formula for how you deal with problems, whether directly relating to your moral stance or not, and it seemed like I was desperately lacking something similar. Whilst maths and science were intrinsic to the development of my own approach to work, these subjects seemed too vast and far removed from each other to be classed under one heading. How could I marry Decision Theory with analysis of Complex Systems or Control Theory?

One thing was clear to me. The need for a coherent approach to problem solving was obvious; if we can improve our approach to work today, we have the possibility of improving our *future*. While there are things we like and dislike about the past and present, the future could have anything we want in it. We just need to apply ourselves in the right way and, with a bit of luck, we can have fame, fortune, time with family and friends, world peace, cures for all known diseases, you name it. It's just a matter of application.

Each of my chosen areas of study helped with this goal. The subjects I'd found interesting hadn't simply provided trivia; they were methods for achieving something bigger. The way I saw it, having knowledge wasn't enough, I wanted new *skills* to help improve the future for myself, for those closest to me, and whomever it was I was working with or for.

Not long after this, I was granted an unexpected opportunity to reflect on these ideas during the national lockdown caused by the Covid-19 pandemic. Being furloughed from my job came as a shock but also gave some perspective; the kind best gained when forced to abandon the stress and ennui of everyday life. One evening, after another trying day helping to home school my three and five-year-old, I sat down at the kitchen table to try and connect the areas I believed had shaped my ideas (rather than lying down in a darkened room as perhaps I should've done). The aim was to take the different disciplines in maths, science and engineering and join them together to create something new: one single, coherent approach to problem solving.

The outcome? This book.

I've spent my career fascinated by the processes we choose to perform and how we decide to perform them. When I started working in Formula 1, I was amazed by the tools that I suddenly had at my disposal and the intellectual prowess of the people around me. The amount of data gathered from the cars at each race was astonishing and the engineers were able to zero-in on their exact area of interest, affording them valuable knowledge regarding the performance of the car. The simulation tools had been built over many years, with huge libraries of car setups that had been run on the track or in simulation. The machine shop was vast and ran 24 hours a day. The wind tunnel was state of the art and constantly churning out variations of car parts to improve the aerodynamics. There was clearly no lack of capability here. However, it often felt like there were big opportunities to improve on the processes that were being used.

The amount of time and resource dedicated to each task could sometimes appear completely disproportionate to the influence it was likely to have. Prioritisation was normally based on the drivers' main complaints during the last race, which often failed to produce anything tangible to deliver, and would therefore turn into a vague look at whatever was making them grumpy. What constituted as 'learning' was often based on only a handful of examples in the telemetry data, with very little exploration into other possibilities.

For me, these indicators suggest we need to improve the way tasks are managed, and not just in Formula 1, but other industries as well. How many times have we found ourselves dedicating our efforts to the wrong things? Can we justify how we *are* spending our time compared to how we *could* be spending it? How do we learn from what has gone before and how do we use this to make improvements in the future? I was convinced that there was an approach that can help us to do better.

It's my view that the relentless march of technological progress has spoilt us in our search for better ways of working. The need to look at our own working practices has been lessened by the availability of ever more powerful machines and software tools. When you sit at your computer or hold your phone in your hand, you are looking at a machine that can perform as many calculations in a second that you as a human being will perform in a lifetime. These devices have more power than all NASA's computers had when they were sending astronauts to the moon and are connected to a network that contains more information than has been available to anyone throughout all human history. And how are we using these incredible tools? Possibly to watch cats being frightened by cucumbers.

The tools we have are just that – *tools* – and therefore we can't expect them to do the work for us. After all, a mallet can't drive a peg into the ground by itself. They can certainly help us to answer our questions, but maybe we should think harder about the questions we *could* be asking. Can we honestly say we've seen the same kind of transformation in our process of managing projects as we have in our gadgets? Our STEM (Science Technology Engineering and Maths) subjects have given us a dizzying array of techniques we could be applying in our problem solving but instead many of us have grown to rely on gut decisions over detailed analysis, experience over skills and short-term results over proper process. We are like children who've been given their first electronic keyboard; it opens the possibility of eventually learning to play Bach, but instead we stick with 'Chopsticks'. We should aim to be more like concert pianists, whose capability is a match for the formidable instrument before them.

In a world as competitive as Formula 1, there are big opportunities for those who can change their practices to overcome obstacles, and this drives a continual search for improvements in the way we operate. Standing still doesn't actually mean standing still; it means a slow trudge towards the back of the grid. Fortunately, many of the problems we encounter are well studied in other industries and there already exists an array of suitable tools available. Over the past few years, my job has been to apply these solutions to the relevant applications.

I have often turned to academia when looking for methods of improvement, occasionally tackling some heavy literature (sometimes literally – certain tomes are useful doorstops around our house), which was far beyond my level of comprehension at the time. In my mind, the ways we will be working in years to come are being studied by academics right now. Surely, if we want to get ahead in whatever industry we are in, we could do a lot worse than look to those with the foresight to consider the future? I like the idea that the research projects we choose to pursue are no different to those that will go on to be published in academic journals. We should therefore expect the same standards. This may seem inefficient in such a fast-paced environment but, over time, can't we develop these methods to produce answers efficiently? That is to say, the only sort of answers I am interested in: valuable, accurate, most of all, *applicable* answers. And get them fast.

I have also been inspired by a favourite past-time of mine: reading a *lot* of popular science. These books tend to go easier on you in passing on information than academic journals do. There are some great communicators in academia who do a brilliant job of summarising ideas from complex subjects and pointing the more curious reader towards more detail where desired. This research started off as a casual interest but, over time, I noticed it beginning to leak into my professional life and eventually start to transform it. Subjects like Experimental Statistics, analysis of non-linear systems and Decision Theory are not taught in any great depth during an engineer's education, and while they offer some very useful practical tools, their real power is in giving a different perspective on the problems we face.

If we are struggling with a particular element of car performance, we are never short of suggestions for how to fix it. As I've already indicated, the engineers in the team are extremely capable and possess a wealth of experience regarding what is likely to work and what isn't. However, it can be a struggle to evaluate these ideas against each other in a way which leads to the best answer. Much of the time the decision boils down to a few laps in the simulator; the driver will give his preference, and someone will set about designing it.

This kind of method has always made me uncomfortable. Sure, the idea we had may have been great for the set of conditions we tested it under, but what about slightly different conditions, different tracks, different weather, a different driver? Have we even completed enough laps to reach a conclusion? Is there anything else we've done in the test which could've biased the outcome? These questions are quite inconvenient when trying to make fast decisions, but that doesn't mean we should ignore them.

While verbally advertising my thinking and my methods, I did manage to twist a certain number of arms. However, some were more comfortable with the status quo. With hindsight, the answer to the questions raised in the meeting over why we needed to consider alternatives was simple; we need to look *beyond* our current capability if we want to avoid the same persistent issues in the future. It's only by doing this that we can truly start to make progress in our activities.

Over the chapters to come, I will demonstrate how we can use ideas from different scientific disciplines to build a single philosophy that can be applied to any industry, any person, any situation life throws at us. This model can explain why we behave the way we do, but more importantly, point us towards optimal behaviours that will maximise the probability of getting exactly what we want, when we want it. I would even go further and say, for me, this explains pretty much everything.

Using this philosophy, we can start to tackle some of the complaints I'm sure we've all had when sitting with our heads in our hands, temporarily defeated by the obstacles littering our paths to success and happiness. We're often exerting effort in the wrong places, neither properly scrutinising evidence nor acknowledging life's uncertainties. In the chapters to come, we will build a powerful new approach to these challenges which will help to smash those obstacles and leave the way clear. It's worked for me. And I believe it will work for you.

2

So, What is it You Want to Do?

What were you doing before you picked up this book? While I'm sure I won't be able to guess precisely, I suspect it was something that you can encapsulate in a word or short phrase, like 'exercising', 'meeting friends, 'cleaning the kitchen' or, who knows? Maybe even 'composing a symphony'. The possibilities are, of course, endless.

But we can probably go further than this. I bet you can describe your entire day as a series of connected 'things' that you have been doing or will try to do. You could start with 'waking up and getting out of bed', through things like 'driving to work', 'reviewing the accounts', 'watching TV', all the way to 'getting back into bed'. We'll usually find the language we use to describe what we're doing is very efficient.

It's the same for all of us, including large groups like project teams, companies, or even governments. Their time can be described as a series of activities, possibly performed in parallel, but always with a purpose, whether business or pleasure, and it is these fundamental activities which we will call 'tasks'.

These tasks can, of course, range enormously in scale. We can take examples like making a cup of tea through to buying a new house. If we take bodies on the scale of governments, we could even talk about huge infrastructure projects or a reorganising of the country's entire health system. These are the building blocks of our lives and we can use them to describe every second of how we spend our time.

In the following chapters we will discuss what all tasks have in common, as well as our experience of how they normally play out. Taking time to consider our problems allows us to effectively derive solutions.

Common Elements of Tasks

Some of the things we can choose to do couldn't be more different from each other. It will be hard to find too many parallels between watching your favourite movie on TV and filling out a tax return. Some we look forward to for weeks, while others we dread for just as long. Take going on holiday compared to your annual appraisal at work. There are however, things that we can pick out that every task will have in common.

All the tasks we will explore in this book involve **human beings** with **agency.** Obviously, everything you have done today involves at least one human being (yourself) but I suspect that in many cases others will have been involved too. Some tasks we will do voluntarily because they give us some sort of satisfaction. Others we perform because somebody with authority has told us to; our bosses, perhaps, or a family member. Even in these cases, it is probably still in our best interests to do them, if only to avoid making life difficult (or having someone else make life difficult for us!).

The success or failure of each task will probably influence our moods a great deal. If we have a good day at work, it will probably mean we're in a good mood in the evening. A cancelled train will mean the opposite. In some cases, the results will influence the moods of others, either directly, like buying a drink for a friend, or indirectly, like buying an electric car to help address the problems caused by climate change. One way or another, we can guarantee that human beings will be involved.

All our tasks will also play out in an **environment**. We tend not to perform them in our own minds (other than 'thinking', of course). We make cups of coffee in the kitchen; we build machines in workshops, and we make repairs to satellites in space. The environments have their own sets of rules and constraints that we must follow to achieve our goals. Getting the most out of the environment we are working in will be key in achieving a successful outcome.

The last thing we will find that all tasks have in common is that we must perform a series of actions or **controls** to accomplish them. This process can be described as a series of steps, performed in a certain order, designed to help us achieve what we want. When we are making coffee, we should probably start by boiling the kettle and putting the coffee granules in our favourite mug. When we are fixing a satellite, leaving the earth is going be an important thing to do at some point down the line.

Because tasks are so fundamental in our lives, I believe we should dedicate some time to understanding why we do what we do and whether we can do it better. The aim of this book is to take these common elements and build a model that we can use to describe any task we might choose to take part in. This could be something as simple as getting ready in the morning, or as complex as running a multi-national organisation. We will call this the 'ACE' model; a model built on the idea that all our tasks will have these things in common: Agent (human or humans), Control, and Environment.

We will be able to use this model in the planning and execution of whatever we have in front of us. We will show which parts of it are captured in each element of the task and how they will help us to apply all essential steps to reach our goal: our problems solved, once and for all.

How Tasks Play Out

Let's walk through an example of how a simple task might play out within our current approach. Your boss has just asked for a report into the sales from last year. They instruct you to stop working on whatever it is you're doing to concentrate solely on this. When you ask about the specifics of what they would like you to cover, the response is something like "just a summary, use your judgement". You immediately start looking at the figures and make some notes.

You spot what you think looks like a pattern, with sales falling by 0.2% early in the year but then rising again by 0.3% later. You make a guess as to why this might be and put it into a report that is around 500 words which covers most of what you've done.

When you present this to your boss, however, they aren't very happy. They were hoping for some comparisons to the previous year's sales and what we might expect to see in the future. Armed with this new information, you head back to your desk and churn out another report. It now looks as though the trend you spotted is unique to last year, but other years have even more pronounced trends. Again, you have a guess at what might have been different about last year and write it all up, with the total word count growing to around 3000 words.

Now your boss is much happier. They can see that you've spotted some interesting patterns and that we should look out for things like this in the future. You have gone into detail on every area of the figures and, while your predictions for the following year seem a little optimistic, they seem reasonable enough. They file a copy of your report in a folder on their computer, probably never to be seen again.

Perhaps you have already recognised some of the problems here. When your boss gives you the task, there is no consideration given to the priority relative to the one you were already half-way through. The aim is not at all clear, engendering the need to repeat the task with clearer instructions. The hypotheses for what caused the near negligible changes in sales, based on only very superficial analysis, were accepted without any scrutiny. Finally, in our long list of grievances, the report is long, boring and will probably never see the light of day again. Laying it out in this way exaggerates these facts, but I bet many of us can recall occasions where we have been through something similar. We can do better than this.

We may well have been taught to look out for mistakes like these in management training courses with titles like 'Nine things to look out for when managing a project'. While we might be able to spot the mistakes in this description and understand why they're important, in my experience, we don't really have an 'umbrella' under which to classify them all.

We are probably all familiar with certain 'heuristics' as part of our daily lives and careers. These are simple rules that you can remember and use to guide you towards a good outcome. For example, "Always document everything you've done for the future", "you can make 80% of the progress with only 20% of the effort", "targets must be 'SMART' (Specific, Measurable, Achievable, Realistic, Timely)". These heuristics are reasonable rules of thumb but in many cases, they exist on their own with no bigger ideas on which to pin them. In my experience, they can grow to resemble religious dogma, particularly when there is more than one idea of how to achieve success. People will divide themselves into tribes, each believing their own set of rules is the correct one to use for the problem encountered.

Motorsport certainly isn't immune to this kind of thinking, despite our best efforts to remain data-driven. "You're not allowed to have a vertical suspension compliance that exceeds 10% of the total compliance". "You can't win at Monza if you don't hit 340 km/h on the main straight". "If you let the mechanical trail grow too large the driver can't feel the self-aligning moment and they won't be able to drive it". In terms of rules of thumb, these aren't bad, but are they solid facts? No. The race stewards won't disqualify you if they find any of these things to be 'wrong' with your car. They are just symptoms of a reductionist way of thinking about car design. This is the idea that if you get every element of your car 'right', you stand a better chance of winning.

It's like this because it is so difficult to get a clear test of anything. By the time the driver has come back into the garage and left again, the track temperature has changed, the driver knows the track a little better and there are more or fewer competitors on track to avoid. These are terrible conditions in which to complete a fair test. When an engineer talks to you about the above topics, it's probably because they've been stung by something in the past. Will they have correctly identified the cause each time? There should be tools we can use that should help us find out.

Even worse is when heuristics allow you to hold mutually contradictory positions. We will often see this when engineers have an idea they want to sell to the people who decide whether it gets made or not. If the part comes out looking good in simulation, "we must trust our simulations when making the decisions". If the part comes out looking as if it makes no difference or worse, that it makes the car slower, it's because "we can't trust our simulations when making the decisions". Seeing this approach to engineering unfold is both funny and heart-breaking at the same time.

We can think of these heuristics as catch-all solutions to problems we don't really understand. When it comes to heuristics surrounding tasks, adhering to all of them in every project is likely to be at best inefficient, and at worst impossible. What we need is something 'Holistic'.

I am a strong advocate of 'Holism', which describes the philosophy of approaching something in its entirety. This is the opposite of 'reductionism', where the whole is considered to be equal to the sum of its parts, and by making all the individual elements better, we will arrive at the best solution. Holism teaches us that everything we deal with will trade with something else. If we want to spend all our money on home improvements, there will be nothing to spend on going out for dinner. If we want to spend all our time investigating what has caused a problem, we'll have no time to implement a solution and if we are spending a little too much time seeing friends at the pub instead of helping around the house, don't expect a warm welcome on your return. Holism teaches us that we are unlikely to attain perfection in everything we do, so we must choose the most important parts and do the best we can. I believe that holism is the only technique guaranteed to put wind behind our sails as we pursue our aims.

By creating a holistic model which can be used to describe any task, we can begin to justify our heuristics. In building the model, we should discover some important ones we have been overlooking up to now. We will also find that some of the rules we have been living by are not 'universal' and we need to think carefully about when we choose to apply them. For these reasons, having a model is a much more powerful approach than simply a collection of disconnected heuristics.

3

Super Models

A term you may have noticed comes up rather frequently in this book is 'model'. We might benefit from taking some time to think about what exactly is meant by this, as it can be used to describe a range of ideas.

A model is a rule or collection of rules that we can use to help describe reality. We can have very simple models, like Hooke's Law, which states that the extension of a spring will be proportional to the force applied to it. Or we can have very complex ones, like mathematical models of the earth that are used in weather prediction. The idea of prediction is key here. We would like to know how something is going to behave in future so we can adjust to it, or better still, manipulate it for our own purposes. When it comes to our task, having models will help us understand how the future is likely to play out before we've even started.

In my career, I've used models of racing cars to predict how making changes to their design affects their performance. These models can be extremely complex and may require a lot of maintenance to ensure they match reality as well as possible. They exist as thousands of lines of computer code which have inputs at the top and outputs at the bottom. In the case of our 'offline' simulations, we are usually interested in a lap time estimation across a certain change whilst using a virtual driver. Again, this 'driver' is just a collection of rules that describes how we think the car will be driven in real life. Creating a complete model of the driver is very difficult, much more difficult than any of the other systems on the car, so in most cases the model we use for the driver is relatively simple. Sometimes the best way to work is to assume a perfect driver and ignore whether the change you have made has caused the car to be more difficult to drive. Other times this is the worst thing you could possibly do!

We can get around the driver modelling problem by using a simulator. Here, the inputs to the model are the same as in the real car, with pedals, steering wheel and other controls. This can be used with a 'driver-in-the-loop' to test how the driver on the track is likely to react to change. Unfortunately, with limited space, we are unable to recreate all the outputs from the car that the driver would normally feel, such as the cornering 'g-forces' and accelerations due to the gradients in the road. The simulator does its best to give the driver the information that it can, which is more than you might expect in the space available.

These models and simulations are heavily based in mathematics, but this needn't be true for all models you might encounter. When we get a driver in to perform correlation exercises between the real car and our model, what are they using to make their comparison? They cannot compare it with reality because they are not driving the real car at the same time, so instead they compare it to an 'internal' model of how the car would behave if they gave the same inputs in real life. This internal model is clearly not reliant on any complex mathematics (drivers tend to be much better at driving than maths) but instead exists in the driver's head as a series of expectations learned from driving the car as often as they have. This comparison is often the most valuable piece of information in the whole correlation exercise, given how finely tuned the drivers are to the cars.

The usefulness of this internal model goes further still. Without it, the driver would not even be able to drive the car. Driving a racing car, or indeed any car, requires knowledge of how it will behave in the future. When choosing when to brake for Turn 1 at the end of the main straight, a driver needs to understand how quickly the car can stop, without needing to test it himself. Brake too late and the driver will find himself in the gravel; too early, and they risk being overtaken by the car behind. Having a good internal model of the car is essential for going faster.

Many of the models we will meet will only exist as internal models in our own heads, or models that are qualitative, with no mathematical derivations. While mathematical models should help us avoid our own prejudices feeding into predictions, we shouldn't necessarily treat these qualitative models as being inferior in the implementation. Sometimes experience with the system of interest can be just as valuable as a complex mathematical description of it. This is particularly true when checking to see if our mathematical models are behaving as we want them to do.

While a lot of time and effort is spent developing, parameterising and maintaining the models we can use, we must all be aware that they will be imperfect in their depiction of reality. They may be incorrect in the expression of the rules that govern the system. They may be incomplete in that they don't consider all the possible ways the system can be affected, or they could be uncertain, behaving one way in some circumstances and another way in slightly different ones. The car models that I use on a day-to-day basis would be virtually useless at simulating a car on a snowy or icy surface, for example, but it's not out of the question that a Formula 1 car can drive on this kind of surface.*

If we have doubts over the accuracy of the model we are using for the task in question, we must improve it. Giving answers that can help us predict how a certain system will behave in the future is the very purpose of the model; if it is not capable of doing this then it is not fit for its purpose. We must be willing to make some 'fudges' to the model such that it behaves as our experiments suggest it should. Note this needn't be the same as modifying measured parameters or violating known laws of nature, as we'll discuss in the environment section.

---

* See https://tinyurl.com/F1icy if you don't believe me!

We must be aware of the limitations of our models, but this is not an excuse to ignore them if they don't give us the answers we expect. Models encapsulate our whole understanding of the problems we are dealing with, so to reject their suggestions is to reject our own knowledge of the problem. This can be best summed up with a quote from George EP Box when using models to make predictions: "All models are wrong, but some are useful".

This 'wrongness' of our model points to another important topic that will come up frequently over the course of our discussions. When we make predictions using our model, we cannot expect them to be perfect. The future will not play out in precisely the way they describe because they are simplifications by their very nature. Instead, we need to think of them as being 'about right', or to put it another way, *uncertain*.

The Concept of Uncertainty

In my opinion, a topic that is not discussed enough in our undergraduate education is the idea of uncertainty. Something that we must become comfortable with is the fact that absolute reality is unknowable and all we have is our own perception of what's going on to guide us. Even with a perfect understanding of the past and present, predicting the future with certainty would still be an impossibility. We should therefore prepare ourselves to meet uncertainty at every step of our journey.

What we mean by uncertainty may change slightly from time to time. We may be able to recall the shapes of classic 'Normal' or 'Gaussian' distributions from our secondary school statistics classes. An example is shown in Figure 1. These will consist of a 'mean' or 'expected' level, with some 'variance', typically indicated by the 'standard deviation'. The variance is our uncertainty of the problem we are dealing with. If you were to choose a random male from the population and measure their height, you would expect it to be around 175 cm, the 'mean' height for men in this country. But as we know, heights can vary significantly between people, meaning we are likely to get anything between 160 or 190 cm, or, quite possibly, values beyond these bounds.

When modelling we must account for parameters that we do not have complete certainty in, as this will lead us to a more robust solution. To get the most out of our processes, we will need to break away from many of the methods we learned in school, as they rely on formulae that work with fixed and precise values. We must begin to appreciate what happens if we have values that are slightly different to what we originally intended – do we carry on more or less as normal, or does a slight deviation bring the whole process crashing down?

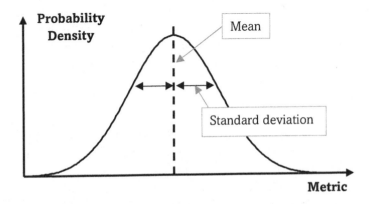

Figure 1. The Normal Probability Distribution

We will learn how uncertainty is hard-wired into many systems, and that the mechanisms behind them guarantee their unpredictable nature. Our universe is not deterministic over long periods of time, and we will not be able to infer how the future will look from only looking at the past and present.

In our society, those who can give us certainties when asked complicated questions are followed to the ends of the Earth. Religion is followed in all corners of the world, with stories that explain all aspects of our lives. Quack scientists who claim to explain the causes of autism in children are listened to intently, and Oracles who profess to have the ability to see the future are paid handsomely by those who do not believe they have equal abilities. Uncertainty is seen as a weakness.

This is not something that will be promoted here. For me, stating your uncertainties in answers you give is as important as the answers themselves. It's no good claiming to be certain when certainty is virtually impossible in most of the tasks we come across daily. After all, it doesn't take a particularly long search of the internet to find examples of those who have expressed certainty in particular ideas only to be proved wrong in quite spectacular ways. The example of the hurricane in the UK being downplayed by veteran weather forecaster Michael Fish is one that is often brought up; the certainty expressed in the report unfortunately didn't stop the hurricane making land fall that night and causing extensive damage.

4

The ACE Model for Tasks

The ACE model was born from my desire to transform my approach to tasks; whether problematic or welcome, whether at work or at home. Over years of honing my skills professionally and nurturing my wider ideas (by way of much extra-curricular research), I sensed my thoughts beginning to link together and form something not only tangible, but *useful*. Harnessing this bunch of ideas and molding them into the ACE model has led to a great deal of success in my career and I want it to do the same for you.

The model consists of four main elements. These are:

- The Agent – the individual or group that owns the task
- The Environment – the systems and processes that are part of the task
- The Control Process – the actions that we will perform to lead us to the best outcome
- Learning and Optimisation – the process we use to learn and improve how we perform the task

I hope you agree that these concepts are simple enough. I intend to explain that their contents and relationship with each other can help us immensely in planning, executing and reviewing any task we choose to perform. We will come to see how these elements are common to every task we can set out on, making this model universally applicable.

While it may be possible to formally state the content of the model, this is not a quantitative model of the kind we might choose to use in simulations. Instead, I expect we will use the ideas in the model to guide our approach to the problems by encouraging a way of thinking beyond our normal patterns. No doubt elements of the model could be used for numerical simulation purposes, particularly when we look at the environment and optimisation of our control process, however, its power needn't be restricted to this type of problem.

About This Book

I have chosen this as a medium of publishing these ideas so that this book can be available to those whom I think it can benefit most. While titles that share the same shelf in the book shop are written from an academic standpoint, this comes very much from the perspective of someone in industry. I hope this means the ideas can be transferred into practical application and not simply exist in an academic journal as a utopian vision for how we should behave. I am not an academic pushing to increase humanity's understanding in any of the areas we will discuss, and as someone who only needs to understand the concepts in order to apply them, I can cover far more ground. For this reason, you may find yourself itching for more detailed information than I can present in these pages. Fortunately, the academics who originally worked on these ideas have produced many of their own books and papers that I will encourage you to seek out wherever necessary.

For me, the best books I have read work through methods to help the reader apply their ideas. They are like practical handbooks with an array of tools that you can use during the task you are performing. I can see how learning about the history of different scientists and mathematicians is of interest to many people, but I am personally more interested in the ideas themselves. I don't really care that Newton and Leibniz fought to the end of their lives over who was the first to discover calculus, I just want to know whether I can use it for what I'm working on. I hope to follow this path in this book. My intention is that by the end you will be able to see the world through these ideas and apply them to your own life and work.

In our discussions, we will come across many mathematical and scientific ideas. The aim of the discussion is not so you can pass exams on these subjects, but instead that you see them as alternative ways of thinking about the problems you encounter. Indeed, my experience is that even those with a scientific background will rarely sit down with pen and paper and express plans with a mathematical framework (there simply isn't time to do this for every decision we need to make). Instead, we can be mindful of the ideas that mathematics and science teach us and use them to create a new set of heuristics to live by. The difference is that these will not be heuristics on solutions to the problems we encounter, but rather the process we go about to find the solutions. These are ideas like Complex Systems and chaos, measurement uncertainty and optimisation techniques. There is enough literature to fill libraries on each of these subjects, but the central ideas can be summarized in a few paragraphs.

I have tried to avoid the use of equations throughout, as I don't feel they contribute a great deal to the discussion on their own. Besides, this is not intended to be a reference book that you dip in and out of. In my experience, it's quite hard to interpret what an equation is telling you in all but the simplest cases. Fortunately, we do not need the equations themselves to understand what they tell us about our reality and how we should act to get the best out of it.

Wherever possible, I will put new terms in single quotes and try to define them. In most cases, these will be common across many fields and it should be possible to conduct more research into them for yourself. I hope you will find a lot of diversity in the topics we will cover, and that this might lead you into areas that you haven't read into before. Because we will cover a lot of subjects, we will only be able to touch on each. I will try to provide references to texts that can provide more in-depth analysis where appropriate.

What Lies Ahead

We will use the diagram in Figure 2 to help us navigate the elements of our model. This will be shown at the start of each section and will indicate how the section content fits into the wider context. We will step through each element in turn, according to the numbers on the diagram, giving time to discuss its importance and how appreciating the principles it captures can benefit us in our future endeavors. It's only when we understand the contents of each of these component parts that we will step back and contemplate the entire model.

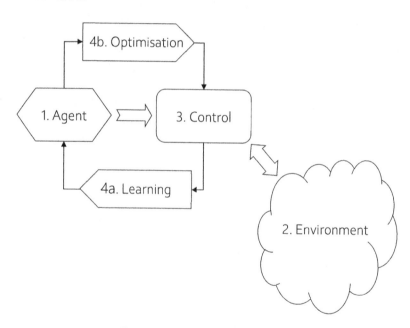

Figure 2. An outline of the ACE Model

Our section on Agents will cover all that is relevant to understanding the preferences of human beings in relation to the tasks they are carrying out. We will introduce the concept of utility and explain how we can map any experience the agent has onto this metric. All agents will have resources that they are able to allocate to the task and these will typically come in different forms, which we will introduce and discuss. While we like to consider ourselves as rational in our decision making, there are certain behaviours that human beings exhibit that mean we cannot always consider ourselves as such, and we're not just talking about the way you might have voted in a recent election... This will be covered under the topic of Behavioural Economics.

The Environment in which the task plays out is full of different systems that we will be able to interact with. In this section, we will show that the world is complex and how this can hinder our search for the perfect model. This will lead us to the concept of chaos and how it affects our power to make predictions about the future. Finally, we will investigate environments with other active agents and how this can lead to non-intuitive results that are explained by Game Theory.

The Control Process will be that which links our agent and our environment. In this section, we will see how we can take the targets given to us by the agent and turn these into actions we make on the environment. Our environment will in turn give us outputs that we can evaluate through our own measurements. We will discuss the types of control process that we could choose to build for our tasks, covering simple discrete steps through to some of the basic ideas behind control theory. We can also consider the process of experimentation here, and how this can lead to a better outcome through a greater understanding of our problem.

Learning and Optimisation are grouped into our final model element. We will introduce the idea of learning from our inferences and see how our understanding can be moved by the results we observe. The different approaches to optimisation will be introduced in this section and we will discuss how we might go about improving our control process to give us a better outcome of the task we are considering. Finally, we will touch on the idea of exploration Vs exploitation, where we must decide whether to stick with what we know or branch off into the unknown.

Following the introduction of each element, we can then bring these together to build the ACE model, thus demonstrating how each element interacts with the others around it. We will discuss how our learning can affect everything from the targets we set through to how we interpret the results. This will lead on to some examples of how the model can be applied to both everyday scenarios and more complex processes which involve many different individuals. We can then show how it is possible to describe the world as an interconnected system of these tasks, from the smallest to the largest scales.

Finally, we will discuss ideas for the future. Where does this way of thinking lead and are there things we should be working towards that we have overlooked since the dawn of civilisation?

The whole process of developing, documenting, and perfecting my approach in readiness to unleash it upon a wider audience has been extremely rewarding, and I hope in sharing it you may be influenced to think differently about the tasks you perform every day.

Let's get started...

Part One

Agent

5

Who'd Want a Problem Like Formula 1?

If you were a company hoping to get involved in a mainstream sport, you'll probably find your options are pretty limited. You might be able to get your logo displayed on a board at the side of the pitch or on a team's shirt, but it will do little to affect the outcome of the match. These sports are typically performed by extremely talented athletes, and while they might need someone to produce equipment for them, the differences between their options are probably small enough that they can't be considered as performance differentiators. Why would we want it any other way? Golf and tennis wouldn't attract the audiences they do if the winner was simply the competitor who could afford the most expensive bats.

In Formula 1 things are different. Here, the equipment (car) that each driver uses is part of the competition itself. There's even a constructors' championship, which ranks the total points scored by each team's car that season. Now any outside company needn't be content with cheering on from the sidelines; they can actually produce a machine that will virtually dictate their drivers' finishing positions. Yes, there are obviously drivers that are better than others but the likelihood of the best driver winning the championship in a below average car is vanishingly small.

You would imagine that those with the most to gain from this arrangement are the automotive manufacturers, who are always desperate to demonstrate their technology and know-how is superior to their competitors. And yes, the sport certainly has these, but they are far from the only brands that participate.

If we look at the grid of the 2022 Formula 1 season, there is an established automotive brand leading the way, but the team expected to be their closest challenger is an energy drink manufacturer. Behind them in the mid-field is a machine tools maker and a fashion label. There is even a team owned by a billionaire from the clothing industry whose son just happens to be one of the drivers. Looking back through the history of the sport, teams have come from all walks of life. So, what is it they hope to gain?

Certainly, having 400 million people watching a car with your company's name on it for ninety minutes every other weekend will help, but there is also the opportunity to associate yourself with other big brands. Teams like Ferrari and McLaren have launched their whole businesses from this platform and now have valuations worth billions of dollars. It might be that by attending the races, you end up rubbing shoulders with members of the global elite in the harbour in Monaco. One way or another, entering the sport could fit your needs and be worth your investment.

For the drivers, the decision is more straight-forward. This sport represents the pinnacle of the single-seater ladder. Ask any young karter what their dream is, and they're unlikely to say "driving in the Swedish Formula Ford Championship". Formula 1 cars are the fastest road course cars that the world has ever seen, and I can't help suspecting that most racing drivers who claim they aren't interested in driving one are lying (though that may be my life-long passion for the sport talking!).

For engineers, the sport offers an opportunity to work with the best people, tools and technology in the automotive industry. The pace is ultra-fast; the record I've seen for having a system conceived, designed, manufactured, tested in the lab, shipped to the track and run on the car is *24 hours*. And the track wasn't even in the same country as the team's base! Again, working in this kind of environment makes you want to get up in the morning.

Central to all of this is the idea of competition; a team that wins the championship has bragging rights over all the others, for one year at least. Your name will be forever recorded in the history books as a champion and that is something a lot of people crave quite strongly. These are just some of the reasons that human beings choose to get involved in a sport like this. In this section, we will discuss the processes they might have used to get them there.

6

Flushing Out the Agent

When we think of the tasks that we come across in our careers and other areas of our lives, the major stake holders are always human beings. It may be that we are considering a task that only affects ourselves, or we could be considering something that will impact millions of people across the world. The fact remains that human beings will be involved.

This is typically in both the execution and in the outcome of the task, although it is not always obvious how. Some tasks, like volunteering at a homeless shelter or working as a doctor in a hospital will have obvious beneficiaries, but the link back to the human beings that benefit from the task may be somewhat contrived; for instance, it is very common for employees in large commercial organisations to complain that they do not see the relevance of the work that they are doing. This is more likely to be a failure in communication from their management than that the task actually being irrelevant - it should always be possible to create the link back to those benefitting. If in doubt, the company's shareholders are a reasonably safe bet, since every time you are asked to check through the department accounts for the third time, you are probably buying somebody another few minutes of time on a beach somewhere with a cocktail...

Because we can guarantee that human beings are involved in all the examples we can conceive of, it is necessary to understand human behaviour and preferences so that these aspects can be included into our whole system model. For this, we will look to the field of economics.

I find length of time it took for me to come across the ideas in this section very interesting; the fact that something as fundamental to our lives as how human beings make decisions should get no mention throughout my entire education. Even in industry there is very little attention paid to the analysis of decision making. While engineers mostly concern themselves with the nuts and bolts of the tasks, without much need for a thorough understanding of human behaviour, there are still plenty of decisions to be made. Some of these are made without knowing the outcome with certainty. In others, we will have no idea of how likely we are to get a successful outcome. Surely greater understanding of these types of decisions will enable us to make more informed choices, giving us a greater chance of success?

Like many, I was indoctrinated through Daniel Kahneman's book 'Thinking Fast and Slow', which paved to way for me to research Decision Theory more thoroughly. This subject is typically split into two parts, the 'rational', economic model and the 'irrational', behavioural economic model. The first describes how we should behave in decision making scenarios, while the second describes how we actually behave. Unsurprisingly, there are some key differences between the two.

Who is the Agent?

We start with a discussion of what we will refer to as 'agents'. This term is used frequently in economics to describe a decision maker in a particular market or model. We will be using this term to describe an individual or group who have ownership over the task being modelled (we will consider how those affected by the task but not involved in its execution fit in later). We will restrict our discussion to human beings in groups of one or more. These are individuals that behave in a predictable, but not necessarily rational, way. Some examples could be:

- A single individual performing a project as part of their job
- A project team
- A department within a company
- A company as a whole
- A group of companies that share a common interest
- An entire country

Clearly this is a very broad definition, and it is intended to be. The final model should be applicable to any area where a group behaves *as one*. The key criterion is that the group being considered shares a common goal but this only needs to be true for the task being considered. A volunteer group that maintains local greenspace can include people from all walks of life, be it employed, unemployed or retired, but all will share the same goal of helping to improve the local environment. This means they can be considered as a single agent in our model.

In my work, the agent of a task can change depending on the work being carried out. It could be my boss (or anyone higher up the company hierarchy) if it's something that will benefit the department I work in. It could be a task that I would like to conduct myself, where I can be considered the agent. Maybe organising some documents into a logical order? More likely, it is something that is being done on behalf the company shareholders.

When we improve the performance of the car, there are plenty that benefit. The drivers may score a lot of points, stand on the podium and maybe even win some races. The team will get bonuses for their contributions and the team principal will be interviewed on TV, leading to the possibility of landing some 'big money' contract offers to move elsewhere. These individuals will benefit when things go well, but they do not 'own' the task of running the racing team; that falls to the shareholders. The role of the team is to adhere to their preferences; if they want to go for broke and spend all their money on trying to win championships, that's how the team should behave. If they are more pragmatic and understand that teams go through their good spells and bad spells, the budget allocated to the operation will remain reasonably constant. As an employee of the racing team, I should be aiming to follow their preferences for how we should work.

There will be situations where we find that *we* are the agent of the task we are modelling, or are at least part of it. Other times, the agent will be giving us instructions that we must execute. In this second scenario, we would do well to understand the true agent's motivations, but we shouldn't rule out the possibility that we don't completely understand our own either. Hopefully the contents of this section will help us in both scenarios.

Utility

Once we've identified the agent in our task, we need to consider how we can describe their preferences. The decisions that we might need to make during the task will be things like, should we do 'x' or 'y'? Both options will have their advantages and their disadvantages, but we expect one to be preferred to the other. What is needed is a scale that we can rank the outcomes of the decisions on. For this we turn to the concept of 'utility'.

This is a term that is common to both classical and Behavioural Economics as well as Decision Theory, and it describes the expected satisfaction the agent will receive from each outcome of a decision. If we receive £100, this is clearly superior to receiving £10, making the utility assigned to £100 higher. Similarly, receiving thanks for a job well done is preferable to disciplinary action for frequent under-achievement. The preferences are likely to be unique to the agent in each case, for example one person might prefer a lot of time with their family over fame and fortune. For other agents, this won't be true, meaning their utility scores for these results will differ.

This fits in very nicely with our holistic approach, as we have taken all the elements of our outcome and compressed them into a single number, which means we no longer need to worry about making trades across multiple parameters when making decisions. As someone who works in a field that only seems to care about final championship position, having one number to work with is ideal. In the commercial automotive sector, vehicle engineers must weigh passenger comfort against vehicle handling performance, the amount of technology against car weight, the power of the engine against the fuel economy and so on. Getting the balance between each of these things right requires a thorough under-standing of the customer base. It sounds like a nightmare compared to just being tasked with making a car go as fast as possible.

Content of This Section

In our model, agents will consist of the following features:

- Utility Function
- Resource

These features could simply be described as what the agent wants and what they have at their disposal to achieve it. We will call the combination of these two elements at any one point in time the 'state' of the agent.

We will first discuss the agent's Utility Function, which will help us to understand their preferences and therefore what we need to aim for when carrying out the task we are studying.

The resource is what the agent can dedicate to the task being considered. This is not only the material or financial resources available, but also the skills and knowledge of human beings in their control, and the time available to them.

We will briefly touch on the importance of the 'state' of the agent in the final model and how it will be used.

As I mentioned in the introduction, we will consider the uncertainty in the individual parameters or functions being described throughout this book. This is essential for ensuring the decisions we make are 'robust'. This concept is relevant to the agent's Utility Function, resource and state, so we'll dedicate time to it in each of these chapters.

7

Fully Functional

What is it that gives us satisfaction when we perform tasks in our daily lives? Are we hoping for fame, fortune, spending time with close friends and family? In many ways, the means of achieving satisfaction are not important. It's the satisfaction itself that we are seeking. This is what we we'll think about in this chapter.

When we perform a task for ourselves, we do it because we believe it will improve our lives (or at least prevent making it worse). When we are asked to perform a task for someone else, as part of a job or otherwise, we presume it will be increasing net satisfaction for whoever requires the task to be completed. In this case, we cannot be guided by our own preferences. We will need an approximation of what gives satisfaction to the owner of the task. This is what we call the 'Utility Function'.

A Utility Function is something that can transform any experience that the agent will encounter into a value of utility for the agent. With a complete function, it should be possible to predict how the agent would behave, or want others to behave towards them, in any scenario. Clearly such a complete function would be horrendously complex and probably not worth thinking about any further. We will concentrate instead on partial Utility Functions that relate directly to the tasks we are performing.

As for *why* we prefer the things we do over others? This isn't something we can concern ourselves with. Clearly there are benefits to society if we can make people's preferences more selfless, like having a greater concern for the environment or the wellbeing of others, but we will struggle to truly change their minds. I'm sure we've all had occasions where we've tried to persuade friends that certain films/music/TV shows are better or worse than they think they are, without success.

While we may at times have ignored the views of others, you have probably noticed your preferences changing over time. I'm sure we've all looked at old photos of ourselves from the past and wondered what it was that possessed us to choose the outfit we were wearing that day? We will not necessarily be able to justify our preferences objectively, so we will not stop to consider why some people prefer chocolate ice cream over strawberry, pop music over classical music, one piece of art over another or anything else you could think of. As someone that enjoys seeking out the most terrifying films in the horror genre, I'm very aware of how bizarre individual tastes can look from the outside!

A 'rational' agent is one that will always seek to maximise their utility in any situation. When picking what music to listen to in the car, they'll choose what they most feel like listening to at the time. When choosing what to do on the weekend, they'll pick their favourite activity, perhaps one they haven't done for a while. When deciding how to invest their time at work, they will choose the project that they think will lead to the most success. This is clearly a valuable starting point for any task that we will be modelling, particularly when we are carrying it out on behalf of another agent. Now we can assume that achieving a result that maximises the agent's utility is the best possible outcome. Once again, we cannot always treat human beings as rational agents, and this will add some uncertainty to our decision making.

We will use the agent's Utility Function to define what a successful outcome of the task looks like; this function will describe the agent's preferences for parameters relating to the task and could include things like:

- Whether the agent would prefer the task to be completed in less time or at lower cost
- The aesthetic form of a product or its functional capability
- Whether the agent is risk averse or more risk-seeking

With this information, we can start to think about what our targets for the task might be. If the agent has a strong preference for getting it out of the way as quickly as possible, it will be no good spending a lot of time coming up with a detailed plan and studying lots of alternatives. Similarly, if the agent is very sensitive to financial loss, spending a lot of money should be avoided. If the agent is not against taking risks with their money, we can bear this in mind when assessing the likelihood of success of a particular course of action.

Before we can do any of this however, we need to establish what their Utility Function might look like.

Appearance of Utility Functions

Consider the wager below:

Would you risk spending £10 on a bet with 50% chance of winning £20 and a 50% chance of winning nothing?

If you're like most people, you'd probably say no. This is interesting, isn't it? Because in mathematical terms, both outcomes, retaining your £10 of a 50% chance of receiving £20, are virtually indistinguishable. If you were a robot, you would be completely indifferent between the two.

This is a good example of how human beings typically behave in the face of risk. This behaviour is well understood and is well documented. Human beings tend to be very wary of this type of wager in a way that cannot be explained by the maths alone. We can use a model to describe how your brain works by way of a simple graph with sums of money on the horizontal axis and utility on the vertical axis.

Figure 3 shows a function with a positive gradient in utility over all values of money (the question of whether an agent can ever have too much money will be left for another day). The curve is s-shaped with gradients that reduce as you get further away from zero in both directions. This reduction in gradient is known in economics as the 'diminishing marginal utility' of money. All this describes is that human beings would be expected to place more value on the first £1000 they receive than the second £1000 when making decisions about money.

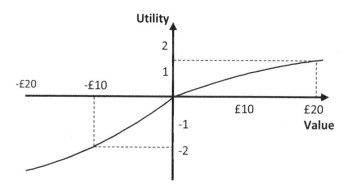

Figure 3. Example Utility Function for sums of money

This is intuitive when we look back through our lives. I'm sure we've all had occasions where money has been tight. Under these circumstances, you probably felt you had to be more careful making financial decisions than in the times you were feeling more affluent. When deciding whether to spend £20 on some new clothes, it's much easier when you have £1000 in the bank than when you have £20.

In this model, there is typically a discontinuity around zero where there is a step reduction in gradient. There is a lower gradient for positive values than for negative ones. This discontinuity is caused because our feelings towards money are not symmetric for gains and losses. Human beings tend to be 'loss-averse' meaning a loss of £1 is likely to hurt more than a gain of £1 will bring satisfaction.

Again, there should be examples from our lives that we can draw on to understand this effect. Those of us who have received parking fines for misjudging the time you had left your car in a car park will understand the pain of having to hand over a sum of money when you've received no tangible benefit yourself. I'm sure this kind of experience sticks in the memory for a lot longer than when receiving a similar amount as a gift. The same can probably be said for any of us that have had both good and bad experiences when investing money in stocks and shares.

Now we have the first element of our Utility Function, how can we use it to predict how this agent would behave in a certain scenario? Let's consider wager that we opened this topic with: would this agent risk spending £10 on a bet with 50% chance of winning £20 and a 50% chance of receiving nothing?

Looking at the shape of the plot, the agent that this represents would not take this bet, as the utility lost for losing the initial £10 stake is much greater than the sum of utilities for £20 and £0, divided by two (as there is a 50% chance of receiving each). This answer is typical for most human beings. Why risk more pain for a chance of a smaller gain when you can stay where you started?

Something to bear in mind is how the shape of the curve will vary across individuals. Different people have different tolerances for risk, with some happy to invest in stocks and shares despite the possibility of losing large proportions of their savings if there is a downturn, while others prefer lower interest savings accounts with the certainty of retaining at least what they invested. We may even come across people who could be considered as 'risk-seeking'; the likelihood of breaking even when playing a national lottery or on casino roulette is very small, however there are those who enjoy making bets on these frequently (and healthily). This is likely to come from the thrill of taking a risk, rather than any expected return on their investment.

While the shapes of these curves will vary depending on who you're asking, the scale is likely to also. The 'value' of £1 to any one individual is likely to change significantly depending on their wealth (and personality) at that point in time. We can refer to our example of the diminishing marginal utility of money again here. Billionaires are likely to value £10 much less than someone on unemployment benefits, meaning they are likely to behave differently when faced with decisions of that magnitude. This causes problems when trying to compare between two agents' utility when only using money.

The usefulness of this plot is that we now have a unique transformation between utility and money. We can take any sum of money, like £20, and discover how much this is worth in terms of utility, which makes comparisons for other experiences much simpler since human beings will find it much easier to relate experiences to sums of money than an ambiguous 'utility' metric. If we want to know what someone considers an experience to be worth in terms of utility, and we have already parameterised this model for them, we can simply ask what is the most they would spend on the experience in question? We can assume that the value of the experience in terms of utility is the same as the value of utility for losing the sum of money they would spend.

With this transformation, we can begin to consider how our agent would value other commodities relative to just money. When we are out shopping, we are making a trade. Do we value the item on sale more than we value the amount of money that it costs? After all, we might find something we like better to spend the money on, or we may prefer to have the money in our savings for a feeling of added security should something unforeseen happen in future.

We are making these comparisons all the time; we will even compare commodities that have no financial cost but instead only cost our time, like choosing between spending time with friends or with our families. What we need is a transformation between money and all other experiences, based on quantities that give us equal utility.

We need to start introducing the commodities we are trading between into our model. For this we can use another type of plot, known as an 'indifference curve'. An example of this type of curve is shown in Figure 4. Here, we take quantities of two commodities and draw contours for how one trades with each. Every point that lies on each contour considered equivalent in the eyes of the agent. For different values of utility, we can draw different contours.

This can help to show the extent to which different commodities or experiences are substitutes or compliments for each other in the eyes of the agent. For example, you could imagine the indifference curve for two famous brands of cola being reasonably straight and a constant gradient, i.e. if your quantity of one were to reduce, it would be completely compensated by the same increase in the quantity of the other. This true for substitutes but not compliments. If you were to consider the same example but for quantities of gin and tonic water, the plot would look more like a right angle, with lines running parallel to each axis. If you started at the apex of each curve (the drink at your ideal strength) receiving additional gin or additional tonic in isolation wouldn't increase your utility, since adding this to your drink would detract from the flavour (in this case you wouldn't add any even if you had the opportunity to have more). Examples of indifference curves for good compliments and good substitutes are shown in Figure 5.

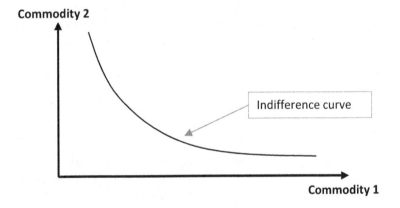

Figure 4. Indifference Curve Example

Clearly these preferences are going to move around over time. We must bear in mind that the utility and indifference curves we have produced are only valid for the current reference point. If things were to change significantly in the future, for example if the agent in question were to suddenly gain or lose a large proportion of their wealth, it will be necessary to re-evaluate the indifference curve starting from this new reference point. If we have a Utility Function for our employer that we have been working to for a while, we should probably check that it still valid when they report record profits for a particular time period. The value they place on money has probably changed somewhat. The idea of a reference point is key to 'Prospect Theory', which we will come onto shortly.

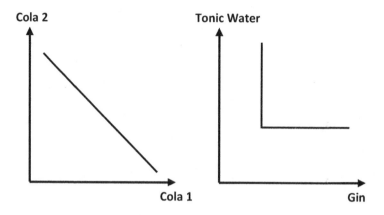

Figure 5. Examples of perfect 'Substitutes' and 'Compliments'

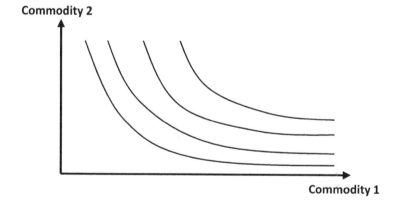

Figure 6. Example of multiple indifference curves

All the above examples we have discussed consider commodities on 'continuous' scales, that is, scales that can take any integer or decimal value. It's also possible to discuss utility gains and losses on discrete scales, for example the expected utility for having one extra car or employee. In these instances, it is nonsensical to talk in terms of fractions of a car or human being, so the discrete scale makes more sense. The utility scale will still work in the same way for these examples, with any option assigned a value of utility. The only thing to bear in mind is that we are constrained within the methods we can use for analysis, relative to the continuous scale.

With these tools, we can start to describe Utility Functions for the agents in our tasks. The team I work for will weight gains and losses of financial resource using a curve similar to that shown in Figure 3. Of course, it is unlikely to be presented in this way, but it will exist in the minds of the shareholders and those at the top of the company. When faced with decisions about how to spend their money or make trades between commodities, they can consult this, along with their preferences for how the relevant commodities trade (their indifference curves).

You can imagine decisions that might affect both finance and the final championship position for the season working like this. Is spending another £1 million worth a slightly higher position in the championship? The answer will come down to their personal preferences. Somewhere there will a limit on what the shareholders are prepared to spend to make the car one tenth of a second per lap faster. Hopefully we now have some tools that can help us with these decisions.

While we have covered how we expect these relationships to look, we haven't really discussed how they can be created practically. If you're interested in trying to create some of these curves for yourself, I have added a guide to the appendix of this chapter.

It's About Time Too

My five-year-old son is sitting at the kitchen table, eyes locked onto a marshmallow that sits on a small plate in the middle of it. There is look of pain on his face as he tries to resist reaching over, picking it up and putting it in his mouth. This is taking all his concentration and will power to achieve. All that's going through is head is that if he can wait just a few more moments, he will be able to enjoy the same experience twice.

He is participating in the famous 'Marshmallow Test'. In this test, the participant (usually a child) is presented with a marshmallow and told that if they can sit for a few minutes without eating it, they will be allowed to eat two. Clearly having two marshmallows is the preferred outcome but being forced to wait a short period of time to receive them splits the participants into two groups; those who felt unable, or simply preferred, not to wait, thus achieving instant gratification, and those who accepted delayed gratification by seeing out the long minutes.

This seems like harmless fun, but the results may have far-reaching consequences for the direction their lives will take. Subjects that were able to wait for the second marshmallow have been shown to achieve greater success in their later lives, as they have demonstrated the ability to put their future selves ahead of their wants and needs in the present. Clearly this will come in handy when choosing how to spend their time. If you can wait a bit longer to start receiving a salary after you finish school, one of several options would be to spend the time in further education learning how to make more when you've finished. If you can put in some hard time at the gym, you may be able to enjoy eating without worrying about your figure. If you can put in your overtime during the week, there's less chance of you needing to be in at the weekend.

The benefits of being able to delay gratification is an example of how the time dimension can be significant when we think about decisions. Humans will typically find it harder to make decisions with consequences that affect us far into the future, rather than receiving them almost instantaneously. Are you saving the right amount of money for your retirement? Would you prefer £10 to spend now or £20 in one year's time? It's difficult to know without any understanding of how our world will look when we get there.

Decisions will not always boil down to such simple trades. On the one hand, a larger financial gain in the long term is still a larger gain (even if you have slightly less time to spend it) but a smaller gain now is so much easier to evaluate in terms of its expected utility. Yes, there will be occasions where we sacrifice a very tangible longer-term improvement through procrastinating, but surely forcing ourselves to wait for good things can't be justifiable in every situation? One argument to postpone a purchase related to your favourite hobby is that in a year's time you will have gained some interest on that money in your bank account, and you might be able to afford something 0.5% better.

Is delaying gratification *always* a good thing? I expect there will be times where it is not. Some of us could probably wait a lot longer before leaving education, trying to absorb as much as possible before unleashing ourselves on the world, however there will probably come a point where we are into diminishing returns. The extra time spent delaying is not really compensated by the outcome. In industry, sometimes a little impatience is what's needed to get a project going. Have you ever worked with someone who is happy to keep trying new things, paralyzed by indecision over which of the very similar options to go with? This is not necessarily a positive attribute.

Of course, time spent 'just waiting' has a cost. There's only so much time you can spend sat staring at a marshmallow before getting bored and taking your chances with what you have right in front of you. This transfers over to activities that we normally think of as pleasurable as well. Do you like watching TV? Probably for an hour or two at the end of the day. If you are pulled away by something else halfway through an interesting program, you'd probably be irritated and wish you could have watched it through to the end. Would you still be happy if you were watching it for 4-5 hours? How about 12 hours? Eventually you must surely reach a point where you are bored stiff and would rather be doing anything else.

We can use time to weight the outcomes of the task utility, with weightings changing as time varies. For some activities, we will gain utility as we continue to participate in them, up to a point where we start thinking that time could be better spent. For other activities, like receiving money, a delay is only likely to mean a lower value of utility. A graphical example for how these weightings might change with time is shown in Figure 7, for biases to both instant and delayed gratification.

Utility payback that stretches beyond the expected lifetime of the agent is likely to be weighted at zero. That is not to say that projects with a payback that exceeds the expected *human* lifespan are irrational. While members of a team/government will come and go, the agent in question may continue to survive for generations, only with different members. This enables projects like infrastructure or environment projects that may take decades to complete to be considered rational. We should however consider that the priorities of future generations may change, as this will affect the size of the payback when it eventually arrives.

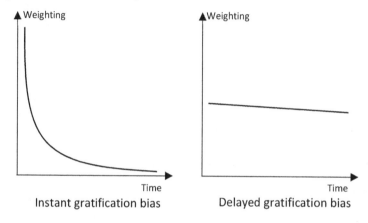

Figure 7. Illustration of different attitudes to delayed gratification

Time clearly complicates things for our task. Some of our choices are separated from their consequences in a medium that we cannot see through, which, unfortunately, is something we're going to have to come to terms with. Maybe we can fill the time by using marshmallows to predict whether our infant children have bright futures ahead of them?

Going Off the Rails

What we have seen up to now are Utility Functions that mostly behave in a nice continuous fashion. Our utility gain between a £10 and £20 is less than the utility gain between £0 and £10 but the transition is smooth and well-behaved. We will come across occasions where this is definitely not the case.

In engineering, we like to deal with 'linear' systems. The behaviour of these is typically predictable and easy to model, for example, a simple electric circuit. Voltage and current vary in a linear way according to simple mathematical laws. Even when systems like this exhibit some non-linear behaviour, say our electrical circuit changes how it performs when it heats up, we usually find that they can be treated as linear by making a few assumptions or simplifications. Now we can use our formidable tool set for linear systems, instead of our very sparse set for non-linear ones.

Alas, sometimes we come across problems where the non-linearities are unavoidable. Racing car tyres are a good example of this. These behave in a nice linear way when they're not working very hard, but things get complicated the closer you get to the maximum grip of the tyre. Of course in motorsport, the *only* thing we're interested in modelling is what things look like at the limit. All our favourite liner system tools are useless to us here and we need to strike out on our own in search of methods of analysis. But it's not only engineering systems that behave in this way; our agent's preferences might end up looking a lot like this as well.

Let's imagine we've booked a holiday that we're very much looking forward to. About a week before the beginning of the holiday we realise to our terror that our passport has expired. This threatens to derail the whole trip. Passport control are not going to be very sympathetic to our situation, no matter how close to the expiry date we are.

Fortunately, we can order an emergency passport. The cost is quite high, but nothing compared to the amount of money already invested in the holiday, all of which we risk losing. We order the passport just in time for it to arrive before the holiday, however there is some uncertainty on when it can be delivered. We have a time in our heads that we must leave the house by to make the flight and if the passport hasn't arrived by then, we may have to cancel the whole thing.

Now we're in a situation where our utility for the passport arriving before our target departure time is very high, pretty much the value of the whole holiday, while our utility for it arriving half an hour later is disastrous. This is *not* a nice smooth function.

Time constraints will feature often in this type of problem, but there are other scales over which non-linear preferences might be found. Think back to your school exams. These are often scored on a continuous scale, but if the exam has a threshold above which you will achieve the qualification and below which you would not, the utility gain between getting that mark and not will be huge. This mark will separate a case of success, with the qualification, celebrations and the possibility of further study, from one of abject failure with huge amounts of time wasted and only the possibility of repeating the exam later. You had better revise!

Looking at examples from my own industry, we can consider the case of big money team sponsorship deals. It's not unheard of for sponsors of motorsport teams to put performance clauses into sponsorship deals. These clauses could cover things like number of times on the podium, race wins or final championship positions. Now the amount of money you receive from this deal isn't only a function of how good a job your negotiators have done, but also on your on-track performance. It's quite possible that your team is on the brink of collapse and has a history of poor results that limits your ability to negotiate large sponsorship deals (teams that run at the back don't spend as much time on TV or get as many 'column inches' in associated press as the more successful teams). You might then have to take contracts that don't pay out much initially but will pay out proportionally more based on your performance. Meet the performance clauses and you can expect more money than you might have got from a simpler arrangement. Miss these targets, and you could find yourself in a more difficult situation than you might otherwise have.

Like the effects of time, these non-linearities are going to cause problems for us and we must be mindful of their existence. In many cases the difference between nearly enough and just enough might be the whole value of the task we are investigating.

What Do You Care About?

The last few headings have described roughly what our Utility Function is going to look like. Now we can consider what it should contain. You would expect this to change significantly depending on what the task we are considering looks like. Are we doing something because we expect the money we earn from it to exceed the time and effort we put in? Are we doing something for the benefit of others? Maybe we're just doing something for the pure pleasure of it.

Whatever our motivations, below are a list of ideas that we might want to consider. Our agent's Utility Function might include all, some or none of these, depending on what we are doing and how we are doing it.

- Financial outcome

This aspect is something that we have followed through from the start. In a capitalist society, the financial outcome of a task (be it gains from an investment or losses because we have spent money on a product with no financial return) is going to have a significant impact on the agent's feelings towards the outcome. We don't need to be fantastically greedy to see the benefits money can bring. Added security, greater luxury and a wider range of options are unlikely to hold us back in the search for happiness.

- Time taken

The length of time required to complete the task is probably something that the agent will have in their minds from the outset. While having the correct outcome in a longer timeframe is likely to be better than not achieving the correct outcome at all (think arriving late for work compared to a train cancellation that means you never arrive), the amount of time taken is likely to have a strong impact on the agent's feelings towards the success of the project outcome.

- Impact on the people affected by the result

Some projects are likely to affect the utility of people not involved in the execution of the project. For example, when launching a new product into the market, we want those that buy it to believe they received good value for money. Selling a product with promises that are not kept will lead to a bad feeling from customers towards the company selling the product, which has value over and above what the company receives in revenue from the purchases. This is something that the agent will be looking for when evaluating the outcome of a new product-based task. Measurement of this (utility obtained by customers) is likely to be more difficult than simple measurements of the financial impact or time taken, but this does not mean it is a less important consideration for the future.

In the tasks that we perform, there are likely to be 'side-effects' that affect others, but that aren't necessarily a target outcome. In economics these are called 'externalities' and they can be both positive or negative (but we're probably more familiar with the negative kind). The impact that a manufacturing process will have on the environment is not necessarily something we plan directly but it is certainly a consequence of production. If we were instead to design our product using recyclable materials, we will not only protect against the immediate environmental impact when the product is disposed of, but we will also be setting a higher standard in our industry for our competitors to rise to. This could be viewed as a positive externality.

- Impact on the people carrying out the work

Similarly, the effect the task has on those performing the work will also be considered. Did those involved need to work 18-hour days in poor conditions with low pay to meet the targets set? If so, don't be surprised to see a high turnover of staff in your company. This will have costs that are invisible if you're only looking at a balance sheet. In a similar way to the previous point, measurement of employee satisfaction is not trivial like measuring income, but it might still be an important consideration when evaluating the success of a project.

- Increased knowledge in the area being studied

We may be conducting a task purely so we can understand a subject area better. This is the whole purpose of academia and we would be wrong to discount the pursuit of knowledge as a reason for doing something. After all, some of the biggest advances in science have occurred just because someone involved wanted to know the answer.

Increasing our knowledge in a certain area is likely to lead to a competitive advantage in any case. Large companies spend billions on Research and Development every year for precisely this reason. In my experience of dealing with a task of this kind, even if we don't learn something we were aiming for, we will often learn something else just as valuable.

What we expect the Utility Function to give us is a weighting between all the above points and any additional ones that the agent considers important. Does the agent weight employee satisfaction as more important because they have worked with them for a long period of time and consider them friends? Alternatively, is the agent a charitable organization with no desire for financial reward? These are the interesting questions that will make every task modelled in this way unique.

As an example, let's consider you are taking your friends for an evening out. This is not a task that you will evaluate the success of purely based on financial losses alone. Yes, you'd rather the drinks weren't too expensive but surely the quality (and quantity) will factor into your overall enjoyment? Similarly, regarding the place you visited, was it too noisy or disconcertingly empty? Did the conversation with your friends flow easily or was it punctured with awkward silences? These are all things that will affect your perception of how the evening went. Indeed, if you went with a partner, you may end up with completely different feelings because the utility they gained is weighted towards different areas. It is critical that we take a holistic view on every experience we had during the task when we are evaluating the success or failure. To ignore these other elements would not give a clear picture.

By considering a range of possible inputs into the agent's satisfaction, we can see why people might be prepared to engage in some activities that, at face value, look completely irrational. A common example favoured by mathematicians and economists alike is playing the lottery. In terms of the expectation of financial reward, the lottery is a pretty bad investment. Some have called it a 'tax on the poor' or even more unkindly 'a tax on stupidity'. If we take the utility from the gain in finances, things start to look even worse, as the marginal increase in the value of money (in terms of utility) reduces as the prizes increase. Despite this, people participate in national and local lotteries all the time. What else could be gained in this process?

Firstly, there's the feeling of 'winning'. If you win something, be it the jackpot or a smaller amount, you can feel like you beat the game, through your own intelligence (very unlikely) or through perseverance. This feeling may well exceed the value of the amount you've invested playing the game. After all, the amount invested is likely to be reasonably low over the course of a lifetime.

Secondly, there is the anticipation of the draws and watching the draw itself. The enjoyment gained from these is likely to be much more significant than for someone not playing. Maybe you imagine the things you would buy with your winnings; you have discussions with friends and family convincing yourself that this week is going to be 'your week' and so on. Finally, let's not forget that the proceeds from these lotteries often go to very worthy causes. Much of the UK Olympic team's funding comes from the National Lottery but there will also be plenty of local causes that benefit from the investment. Even if you do not win on a given occasion, you can go to bed knowing that your money has done some good.

This line of reasoning can work with other forms of gambling. You may not be donating to a worthy cause, but you are likely to increase your enjoyment of a day at the races by placing a responsible bet or two. Yes, these are likely to be bad investments for those who don't have an enormous amount of knowledge about the sports or games being played but that doesn't mean you are an irrational human being for trying.

Sorry, I'm Not Finished...

Imagine you are choosing a restaurant to eat from on a local high street. We have a choice between Indian, Chinese, Italian or Malaysian food. For most of us, Malaysian food is not something we come across as often as the others, so how are we supposed to compare its value to those of the other options? Our Utility Function for this decision is missing this value.

If we were to look at the menu of this restaurant, we might find that we don't even recognise many of the ingredients. The prices seem reasonable, but we have no idea of the quantity we'll receive with each dish. The safest option is to go with what we know and eat in one of the restaurants we already have experience of. If we are risk-seeking, we may choose to try something new but only if we are prepared to throw away the price of a meal on something we don't enjoy eating.

We should expect our agent's Utility Function to be 'incomplete' in a similar way. Would the agent enjoy kite surfing? Maybe an evening of Japanese Kabuki Theatre? Perhaps a holiday in Mongolia? While it might be possible to make a guess of where our agent's preferences might lie, we're probably going to have to make some big assumptions and these are likely to come with a great deal of uncertainty.

The fact remains that however thorough we are in building our agent's Utility Function and however much we understand our environment and its possibilities, there will be cases that we can't put a value on because we've had no prior experience of them. The world is full of different experiences and there will be some that will be completely alien to us.

In most cases our function will be incomplete by necessity, as creating the whole thing will be far too time consuming an exercise to be rewarding for the task. In other cases, they will be incomplete because we have no idea what our preferences for certain experiences are. If we aren't sure whether we are going to like a particular item on a restaurant menu, better be safe and order the burger and chips like we always do.

The problem is, if we take the safe route every time, we will still be faced with this decision on future occasions. We can probably think of times where we had a weekend free and rather than trying something new, we reverted to tidying the house and watching a film we'd already seen. This cost us a valuable opportunity to go out into the world and learn more about it, and ourselves. We will find out later why this may be more rational than exploiting the solutions we are already familiar with, particularly if we have a lot of time to fill.

With the above headings, we've covered the features we might expect to see in our agent's Utility Function; we can expect our function to be multi-dimensional and have features like loss aversion and diminishing marginal utility for commodities. Time is going to be significant and can be used to describe preferences for instant vs delayed gratification and non-linearities are likely to exist when the things we are interested in come up against discrete thresholds for acceptability. Finally, we can expect our function to be incomplete, with gaps that will prevent us from assessing whether some outcomes are preferred over others.

This function is going to be unique to the agent we are considering, both in terms of the dimensions it is made up of and the preferences it expresses. It may require weighting the preferences of multiple individuals, depending on the type of agent we're dealing with. If we're only considering ourselves, we have access to the complete function, albeit as an internal model rather than something formally expressed. This is no doubt more efficient than trying to estimate the appearance of others' functions, but we should be mindful that we will probably lack certainty over some preferences.

With this understanding, we can begin to think about how we should behave to achieve our desires. The processes we need to attain them are likely to be complicated, however, we can probably break them down into fundamental decisions that will dictate the path we take. It's these that we will start to explore in the next chapter.

8

Decisions, Decisions...

Now we have established definitions for an agent's utility and their Utility Function, we can begin to consider how human beings behave when making decisions. We can think of decisions as opportunities for us to manipulate the future to suit our desires, by choosing from the options laid out in front of us. This discussion can be split under two separate subheadings: rationality, or how we *should* behave when making decisions, and 'Behavioural Economics', or the science of how human beings *actually* behave when faced with decisions. These topics fall under the broader subject of 'Decision Theory'.

We've already mentioned that the key idea when evaluating how to behave when faced with a decision is that we are attempting to *maximise* our utility. Why would we want anything other than the best outcome from a decision? When we are choosing music to listen to in the car, do we choose something we don't really like or aren't in the mood for? When we are deciding how to save our money, do you choose the account that has the second-best interest rate? I'd hazard a guess at "no". If we want to make the most of our time on this planet it makes sense to go for the option likely to bring most satisfaction.

Let's start with a rational approach to decision making. Consider the following wagers:

90% chance of winning £10, 10% chance of winning nothing

10% chance of winning £90, 90% chance of winning nothing

50% chance of winning £12, 50% chance of winning £6

If we assume linearity between sums of money and utility for a moment (£2 is twice as good as £1 etc.) and the probabilities are certain, which of these wagers is best? You may have a preference, but a rational agent would find these options indistinguishable. Why is this? Each of the wagers above has the same 'expected' outcome; winning £9. We can calculate the expected outcome by multiplying the probabilities of each result with their values and adding them together. In the case of the third wager, we can take 50% of £12 (£6) and add 50% of £6 (£3), to give us £9. A little bit of maths should help you to verify that the expected value of the other two wagers is the same.

This expected value is interesting because in none of these wagers is it possible to win £9. The expectation is that, over a lifetime of making similar decisions, the average return you will receive will tend to the expected value of all the wagers. We call this the 'frequentist' interpretation of these wagers.

The frequentist interpretation of probability is one that you will almost certainly have come across in your school maths classes. What's the probability of getting heads in a coin toss? 50%. What's the probability of rolling a 'three' with a standard die? One in six. What's the probability of winning the jackpot in the national lottery? Basically zero. Hopefully this is all sounding very familiar. The idea is that any probability is derived from the proportions of each outcome you would expect to see when the number of trials you perform in a certain task grows to be very large. This way of thinking about probability has given rise to some incredibly useful tools in the field of statistics but there are those that consider it far too restrictive.

In life, we are rarely offered the opportunity to participate in wagers like this an infinite number of times. Indeed, you may only be presented with an important set of wagers once, so what we need is a method that can help us with this kind of decision. Here, we could use the so-called 'maximin' approach, in which rather than looking at the expected values, we look at the worst possible outcome we could get in each of these wagers and choose the one that is the best. This is the 'least-worst' outcome. In the first two wagers, the worst thing that could happen is that we win nothing. Not the end of the world, but not any better than our current situation either. In the final example, the worst that could happen is we end up with a gain of £6, so by taking this wager we have guaranteed ourselves a final state that is better than what we started with. This is the most 'robust' choice, rather than the most rational one; it minimises the pain we could receive by participating in this choice of wagers.

The maximin approach to decision making is going to be useful in other situations too. For example, while the above wagers are accompanied by a nice complete set of probabilities, it's possible that we could be faced with a similar decision with no way of knowing what the probabilities of each outcome are. Note that this is not quite the same as being uncertain over the probabilities (if we know roughly how often different outcomes appear, we can make an approximation and carry on using our expected value approach). For decisions made 'under ignorance', we can't calculate the expected outcomes of each of our options because we're missing key values from the calculation.

These decisions usually arise when you have very little knowledge about the area that you're dealing with. You are unlikely to have experience in every field that requires a decision, much less have a detailed understanding of the probabilities of receiving each of the possible outcomes. Some examples might be:

- What is the probability of drawing a yellow ball out of a bag with unknown contents?
- When lost on a road when out driving, what is the probability that the next left turn will take you to a road you recognize?
- What is the probability that hostile alien life will be found in the next 1000 years?

Again, without probabilities, we can't calculate the expected values of each decision, but our maximin approach still works. If we take the first example in the list above and imagine we would win £10 by drawing a yellow ball from the bag and lose £20 if we don't, we can see that by not participating we avoid the possibility of losing anything. We can think of this kind of decision as the rational approach, rather than just the robust one.

We can use this logic for any similar decision made under ignorance. If I continue driving down a road I don't recognise, without any means of navigation, the probability that I will suddenly arrive somewhere I do recognise (good outcome) is unknown, as is the probability that I will end up lost for hours (very bad). Alternatively, I could stop and ask for directions and at worst I am only a few seconds behind where I would have been (not too bad) and at best I have directions that will get me back to where I want to be (very good). Again, the second option contains the 'least-worst' outcome, namely only being a few seconds behind where I would have been without stopping. This is therefore the rational action in this decision, as your partner may well have already pointed out to you in the past.

The Right Tools for the Job

The examples we've discussed so far involve very simple choices. Do I participate in this wager or not? Should I stop and ask for directions? While their solutions may help us to address simple problems, much of the time we are likely to be dealing with things that are more complicated. We are probably going to have to decide between options that have some advantages and some disadvantages, which we must weight appropriately to reach the solution that's right for us.

My wife and I recently decided to move to a new house. We hadn't really settled where we were living, and we felt that trying somewhere new was a better option than staying put and hoping things would improve. This is clearly a big and complicated decision, something that creates a tremendous amount of upheaval and has big consequences for going the wrong way.

In this decision we were going to have to trade things like the size and design of the house that was up for sale with how much we liked the surrounding area, the quality of the schools that it was close to, the expected length of the commute to work, the likelihood of flooding in the area and many other aspects. It isn't the probabilities that are going to give us a headache here, it's the sheer complexity of the consequences.

In these cases, there are some simple, effective tools we can use to help get us started. We obviously need a reasonable understanding of what our Utility Function for these decisions looks like. In the above example, we should try to understand how much we value the perfect house, good schools etc., but this needn't be in the form of a formal Utility Function with indifference curves split across multiple dimensions. It can simply be enough to know what feels roughly equivalent in your own mind. From here we can introduce the method of 'Even Swaps'.

Say we are deciding between two cars. These are different over many characteristics, meaning it's not obvious which meets our needs best. The Even Swaps technique involves trading characteristics into the value of other characteristics until an obvious favourite becomes clear. If we are deciding between a hybrid car with excellent fuel consumption and a non-hybrid car with more boot space, we can decide how boot space trades with fuel consumption from our perspective, remove one of these dimensions, then correct the other. For example, if we think 10 litres of boot space is worth about 1 mpg and our non-hybrid car has 100 litres more boot space, we can update the fuel economy to make the non-hybrid 10 mpg better and discount the boot space entirely. This is illustrated in the table below.

Car	Fuel Economy	Boot Space
Hybrid	60 mpg	~~300 litres~~
Non-hybrid	40 mpg	~~400 litres~~

→

Car	Fuel Economy
Hybrid	60 mpg
Non-hybrid	50 mpg

Figure 8. Even swaps table for deciding between different cars

Now we've removed the boot space dimension, we can see that our hybrid car comes out as preferred. The difference in fuel economy cannot be completely offset by the difference in boot space between the two cars according to our own preferences.

This method can be used across many different dimensions, provided we are able to trade them against each other. We simply need to eliminate one dimension at a time until we are left with just one. If we want to, we can even use non-linear transformations between dimensions; for example, we are likely to value an increase of 20 horsepower on a 100-horsepower car more than we are on a 500-horsepower car (maybe we can refer to this as the diminishing marginal utility of horsepower?).

What we will be left with is a single dimension that is the sum of the characteristics expressed in the units of the remaining one. We can think of this as a score in utility but expressed in strange units, for example 'mpg' in the example above. Because utility is expressed on an arbitrary scale, this will be of no consequence. If we do have a detailed, formally expressed Utility Function, we can use it to get utility scores for each dimension without having to resort to trading each of them off with each other. We will find that this is a much more reliable approach as it reduces the influence of uncertainty in the trades on the final answer.

The even swaps table can be thought of as a simplified set of Indifference Curves. While these tell us how combinations of commodities trade against each other in the minds of the agent, we can also use them to tell us how to maximise our expected utility. If we are going to buy combinations of these commodities, for example buying gin and tonic water from the shop, we must be mindful not to ignore the cost of each when making decisions. Something we can do to help is to add contours of cost to the indifference curves; this will allow us to subtract the utility of the cost from the utility of the combination of commodities to help with the decision.

Take the plot in Figure 9. With the addition of the price contours, we can begin to see what levels of each commodity will maximise our utility at each price. Looking at the line corresponding to £30, the grey circle indicates our optimal combination of commodities, giving us utility of +1. All the other combinations on this price contour give lower utility than this.

As we vary the price, we can find more optimal points and join them up with the dotted line. If we combine this curve with our utility curve for money, we can now decide exactly how much we should buy if there's no fixed budget; we simply take the utility gained by the optimal combination of commodities and subtract the utility lost from paying the cost. We should be left with a single 'optimum', which is the combination of commodities and the associated price. This is the amount that will bring us greatest satisfaction across the dimensions of each commodity and money.

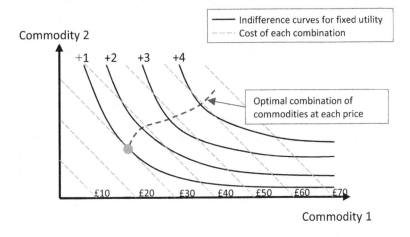

Figure 9. Plotting cost contours on top of indifference curves

Good Prospects

While the above is a brief summary of how we would ideally behave when faced with these kinds of decisions, unfortunately we cannot assume that human beings will always act in such rational ways. In their ground-breaking work, Kahneman and Tversky presented 'Prospect Theory', which helps us to explain how human beings behave when faced with 'decisions under risk' and went on to earn the pair the Nobel Prize for economics in 2002.

Kahneman's book 'Thinking Fast and Slow' is one I can't recommend highly enough. I remember absorbing the whole thing over the course of a very enjoyable summer holiday. The premise is that there are two different thought patterns that we use when evaluating problems. The first is the 'fast' system, which is very light and based on efficient cognitive processes. This is a relic left over from our evolutionary roots in the animal kingdom (It's no good stopping to think what to do when there is a predator pursuing you though the forest). The second system is much cleverer, but this comes with the cost of being slower and consuming a greater amount of energy. This is far more unique to humans and is where all the intellectual heavy-lifting is performed. Because of the inefficiency of this system, we'd rather avoid using it at all, which is why we find ourselves making silly mistakes when presented with problems that are more complex than they appear at first glance.

The first half of the book focusses on how the systems differ in terms of their strengths and weaknesses, and there are some interesting examples of problems caused when the fast system overrules the slow system prematurely. This is the part that most of the people I know that have read it found most enjoyable. For me, the real value comes in the second half, where the theory that brought the pair their greatest success is discussed.

Prospect Theory is divided into two parts. The first part explains that the expected utility for the outcome of a decision is evaluated on possible gains and losses relative to the status-quo, rather than changes in 'wealth' as had always been considered the case before. For example:

You are given £10 - would you take another £5 for certain, or risk it for a 50% chance of obtaining another £10 and a 50% chance of gaining nothing?

You are given £20 - would you take losing £5 for certain or risk a 50% chance of losing £10 and a 50% chance of losing nothing?

In both cases, the options are the same - £15 for certain or a 50% chance of either £10 or £20, but did you find yourself leaning towards accepting the gamble in the second wager but rejecting the first? Here's an example of your fast system drawing a premature conclusion. This is due to differences in the 'reference points' of the two wagers. In the first, the reference point is set at £10 so any additional money is seen as a gain, whereas in the second question, all options are losses. The evaluation over the Utility Function is therefore different in each case.

We can think of the reference point as the position of the status-quo. If you were told that you had won the lottery jackpot one morning, only to find that you would lose most of it because of tax in the afternoon, you are likely to feel less happy than if you had won that smaller amount originally but were told you could keep all of it. This is because our reference point shifts quickly to accommodate changes in our wealth such that each subsequent action is evaluated as a gain or loss relative to this point. The revelation that you will lose a lot of your winnings to tax is a loss relative to your previous state of having just won the lottery jackpot.

The second part of the theory describes how human beings tend to weight very high and very low probabilities differently from what would be expected if they were perfectly rational. Human beings tend to over-estimate their chances of receiving an outcome with a very low probability and under-estimate their chances of receiving an outcome with a very high probability. For example, which of the pairs of wagers in each of the following cases is your favourite?

50% chance to win £650 or a 52% chance to win £630

98% chance to win £650 or a 100% chance to win £630

In the first choice, the second option has the higher expected return (£327.60 compared to £325), while in the second, the first option has the highest expected return (£637 as opposed to £630). Most people will choose the opposite wagers in each case when presented with the same questions.

If we take the second example, a 98% probability is still basically certain when considering the decision in isolation. A rational agent would certainly value the additional £20 more than this slight loss of probability. When we predict how human beings will behave when presented with these wagers, we must apply a 'weighting' to the probabilities of each outcome. This weighting will be above one for low probabilities - as human beings inflate these - and below one at high probabilities where we tend to under-estimate our chances. Human beings tend to do better with positions of certainty (0% and 100% probabilities) and for probabilities close to 50%.

These behaviours should certainly be considered when predicting how individuals will behave in similar situations, or if making decisions on behalf of someone else. If you are unfortunate enough to get the outcome you didn't want after making the rational choice, be prepared for a (possibly patronising) discussion where someone explains to you why you shouldn't have made the decision you did.

This theory has been used to explain many of the phenomena that appear in Behavioural Economics. One such example it the so-called 'Endowment Effect', which helps to explain the difference between the price someone will sell a commodity for and the price they are willing to pay for it. We can think of this a change in reference point; from a state where they don't own the object to a state where they do.

To explain this further, we can refer to an experiment conducted by Ariely over the price of college basketball tickets. These tickets were very strongly desired by many of the fans, but rather than being able to purchase them for a fixed price, they were distributed to winners of a lottery. After the tickets had been allocated, the researchers questioned the winners over how much they would be prepared to sell them for, as well as questioning those without tickets how much they would be prepared to pay for them. It was proposed that anyone willing to pay more than the lowest amount a winner was prepared to sell for could then go and buy the tickets.

Unfortunately, the prices those without tickets were willing to pay came nowhere near the amount the winners would be prepared to sell for. In this case, the disparity was huge, of the order ten times in some cases. Those who had won clearly had significant attachment to the tickets that they owned compared to those who hadn't. Much of the testimony from the losers even suggested that after the draw, they found they didn't want the tickets as much as they had thought initially.

In this example, the participants who went on to win the tickets initially approached ownership from the perspective that it would be a gain. As we know from our utility curve example, gains and losses are not valued symmetrically by most human beings. After owning the tickets for some time, their reference point has shifted to one where no longer having the item would constitute a loss, with a value of utility that far exceeded the gain that they initially received. We can therefore see this effect as a form of loss-aversion.

Prospect Theory has demonstrated that human beings do not necessarily behave rationally when faced with decisions, so it would be wrong to assume this in our model. We are going to have to keep a close eye on reference points as our task progresses, and not get too upset if our project is rejected for investment the day after the company reports a massive drop in profits. This of course makes our problem more complicated, but hopefully in a particularly interesting way!

I Can't Be Certain

We will now switch our attention to sources of uncertainty in the expected utility of our decisions. Even if we have a complete Utility Function to represent the agent for the task in question, we can still be surprised that the outcome is not what we expected. We will consider the following sources of uncertainty here:

- The uncertainty inherent in putting together the Utility Function
- The possibility that the shape of the Utility Function changes during the task
- The possibility that we have overlooked a certain human behaviour when predicting the agent's preferences

Firstly, when compiling our Utility Function, we are taking *measurements* from the agent we are interested in. We will discuss the uncertainty of measurements in more detail in our section concerning the 'Control Process' but here we will say that we are unlikely to find all our answers lying a smooth curve or plane. We will therefore be left to 'fit' our Utility Function through a cloud of points. While we can take certain steps to ensure the function we arrive with is the best possible fit through the data, it will not pass through every point of data we have collected. The difference between the *residuals* (the measurements we have taken minus the curve or plane we have fitted through it) will give us some idea of the level of uncertainty in our model for the function.

What is more likely is that we will only have a vague, informal idea of what the agent's Utility Function looks like. This is true in most of life's simple tasks, for example I don't know the most I would pay for certain products or experiences with much certainty. We are normally faced with simpler yes/no type scenarios when making purchases and coming up with complete utility curves for every decision we make is likely to be more effort that it's worth, particularly when making the decisions for ourselves. Clearly in these cases there is a lot of uncertainty over whether one outcome really is better than another, particularly if you don't get to experience the alternatives.

Prospect Theory shows the reference point of the agent is very important when predicting how they will respond in certain decisions, even for problems with identical outcomes assessed differently according to where the agent has placed themselves as their 'reference'. Alternatively, the reference point may have changed during the execution of the task; how many times have we carried out a job for a boss only for them to change their mind halfway through? This is not necessarily irrational behaviour, as new information could arise in the intervening period. Indeed, information acquired during the completion of the task itself may change the way it is viewed. As an example, if you were a new company that was looking for oil reserves in a promising area but on your first attempt at drilling you came across large quantities of gold, this is not necessarily a reason to try drilling somewhere else...

Whilst Prospect Theory and its consequences are key in the subject of Behavioural Economics, this subject has a great deal more to teach us about other irrational behaviours we participate in without even realising. These will all contribute to the uncertainty we will come up against when predicting how others will make decisions.

The phenomenon of 'Anchoring' describes the tendency for human beings to fixate on the first price that they hear when making decisions over purchases. This 'anchor' will remain in your subconscious for as long as you have experience with that type of product, for example, imagine your TV has just stopped working; how much would you expect to pay for another one? We might be able to get a 28" flat screen for around £250, which is acceptable but not exactly top of the range. We could choose to spend more or less depending on what is important to us. Excellent picture quality, catch-up TV, high contrast screen with vivid colours? Whatever the requirements, there will be a model that represents the best value for money for each of us.

Now let's consider a parallel universe where you cannot get a TV for £250, or even £1000. In this universe, a basic spec will set you back around £3000 (but everything else you can buy is similarly priced). Do you think you would still own a TV? Let's think about what we gain from our TVs. There are estimates that the average American spends up to 4 hours a day watching TV. While it is perhaps less true in today's smartphone generation, in the past the TV has been the main window from each household to the outside world. We use it to get news, watch sport and be entertained. All this for a few hundred pounds. I suspect in this parallel universe we would still own TVs, even though the price seems extortionate when compared to prices we're familiar with.

We need to remember that the price of goods is set by the market, or a combination of supply and demand. TVs are cheap because with today's manufacturing, they are reasonably easy to produce and distribute, not because we do not value them as much as other, more expensive items. When we see TVs costing thousands of pounds, some of us wince because we can get cheaper ones, not because we think they aren't worth that amount. This is one of the effects of 'anchoring'.

Anchoring has been shown to happen very quickly. Ariely showed that the price of experiences is anchored to the first price the customer hears, even for multiple subsequent experiments involving different price ranges.

Experts who pride themselves on being objective have also been shown to be influenced by this effect. In an experiment by Northcraft and Neal, a group of estate agents were asked to value a property that had already been valued by another agent. The participants were split into two groups who were told different results of the initial valuations, one much too high and the other much too low. They were told to ignore these guide prices and give their own opinion about how much the house was worth. Despite being told to ignore the guides, the group that were given the higher price initially valued the property more highly that the group given the lower guide. In their feedback, they were adamant that they hadn't been affected by the price they were given at the start of the viewing. This is extremely unlikely to have been true. (As an aside, we can use this to our advantage in our own negotiations. The party that states the first price sets the tone for the remainder of the negotiation, even if this price is completely unrealistic).

When predicting the decisions others will make, the effect of anchoring will be difficult to account for. We may find that individuals from poorer backgrounds will accept a significantly lower wage than those whose background is more affluent. Are the richer individuals greedier and more arrogant? Not necessarily, they may simply have anchored their expectations of what to expect for a day's worth of work to their previous experience, rather than the price set by the market.

One of the more worrying effects that we should be concerned about is that we might not be very good at predicting what is going to bring us happiness *at all*. In one of the most popular TED talks 'The Surprising Science of Happiness', Dan Gilbert discusses the idea that we human beings are quite bad at anticipating what will make us happy but in other scenarios we are quite good at 'synthesising' happiness (If you haven't seen this video, I recommend you seek it out. Some of the examples given are astonishing!).

It shouldn't surprise us that the things we think we want might not necessarily bring us happiness or utility. There are plenty of self-help books out there that might be able to help us understand our true selves. For the time being we'll carry on with the idea that we broadly understand what it is we want. The alternative would be to aim to satisfy what we believe what our superiors would like, rather than what they've asked of us. I wish anyone going in this direction good luck...

There are plenty of other ideas in this field that show human beings are far from rational in the choices we make; the above ideas are far from an exhaustive list. The main takeaway should be we should not assume the Utility Function we are working with is perfect.

It is bound to be subject to a great deal of uncertainty and we should probably keep track of how things are going along the way.

What Does This Mean for Our Task?

We can now turn to how we can use the above information when modelling our task. The agent's Utility Function describes their preferences for the outcomes of the task; we understand how their utility for money changes over different amounts and what they would consider efficient trades between money and other commodities; we understand how they can delay gratification and whether there are any 'non-linearities' in their preferences. We also know that all the above is uncertain, both because we will struggle to get an accurate Utility Function and because Behavioural Economics suggests there are quirks of human decision making that will affect us as we go.

We now have all that we need to describe the 'targets' for the task we are involved with. We are used to seeing targets like a specification for the task at hand; I want a car with at least a 1.6 litre engine, I want our turnover next quarter to be at least £10 million, we need to win at least 60% of the vote. With our Utility Function, we have something more powerful than this. We know that we can probably still maximise our utility even if we don't meet these specifications, provided we make up the gains in other areas. Take an example where a company wants to break even at the end of the year; they make a small loss, but they end up with a larger market share, an increase in customer satisfaction and loyalty, and an incredibly happy set of employees. Will the still consider the year a failure? Probably not, as they have gained a great deal beyond an over-simplified target for success. The likelihood of them making a larger profit in following years is probably much greater.

With a suitable Utility Function for the task in question, these arbitrary specifications become obsolete. In my mind, this is no bad thing; if we meet a specification, are we incentivised to look further and try to do even better? Not in my experience. If we push as hard as we can after exceeding the target, we will only make our lives more difficult when new targets are set (this is known as 'satisficing'). If the target is instead to maximise the Utility Function *at all times*, we avoid the problem of targets being set that are just too difficult for anyone to achieve, while maximising our return in cases where the targets can be hit easily. Does this mean we shouldn't have any expectations for what might be possible in the task being carried out? Of course not, but we should be trying the best we can in all scenarios, rather than only until we hit a target.

Of course, the same will be true of any deadlines for completing the task. Yes, we should always have an expectation of how long a task should take but if we take a day longer on a 2-year task to ensure we make double our expected profit, is anyone going to be upset? I would think not.

With the information we have gathered, we can advance our task beyond the typical target/deadline mindset into one that appreciates the trades between different outcomes. The targets can now change in real time as a function of the information that we get back from the environment. We can ensure that we always maximise our potential at each point in time, rather than sit back and enjoy having met all our objectives. We're making progress already...

9

Show Me What You've Got

For most of the tasks we encounter, simply having a preference for the outcome won't be enough to attain it. We should probably expect to make some kind of effort. Maybe we'll have to dedicate some of our time or spend some of our money; it may even be that we need some specialist understanding to even attempt whatever we have set our hearts on.

This is going to be the other side of the balance to our outcome. How much effort should we be prepared to make to get what we want? There will be occasions where the effort we would have to make will not be compensated by the result. It might be nice to make Michelin-star quality food for dinner but being able to do so will require many years of training and a significant proportion of each day dedicated towards cooking. For other tasks, the outlay required is basically nothing compared to the utility gained - an evening in front of the TV will require a small amount of electricity and plenty of sitting down.

We will use the term 'resource' to describe what the agent has at their disposal to dedicate towards the task at hand. The entirety of the resource will define the agent's capability; a multi-national organisation with thousands of employees and millions of pounds in their budget is probably going to be able to achieve a lot more than any individual, however, all this resource clearly won't be necessary for every-day, simple tasks.

The resource types that we need to consider will not necessarily be the same for each task we model. We will cover three different types in this chapter. These are listed below:

- Material
- Skills & Knowledge
- Time

We will discuss each of these in turn, considering their importance to the modelling of the system.

Material

We will define 'material resource' as the agent's assets that have a monetary value in the context of the task being undertaken. At the very least, these are things that can be traded for other products or services that will influence the outcome of the task. We might not be able to use our old car as a deposit on a new house, but we could probably sell it and use the money we make instead. Money in the bank is an obvious example of material resource, but there are others: machine tools, software, infrastructure, financial credit, or even contacts the agent has outside the task, who can be called upon for advice. We could use these towards the task to improve the outcome, or we can pay for new non-monetary resources to help accomplish the task ahead of us.

Once we have spent our material resource on the task in question, it cannot be used again, unless we have bought something that will outlive the duration of the task. This should not come as a great surprise to anyone. If we want to mow the lawn one day but we don't own a lawnmower, we will need to buy one. This would be a great expenditure for one job, but the good news is that we will be able to use it many more times in the future. It would therefore be reasonable to spread the cost of the lawnmower over multiple uses. However, this resource can no longer be spent on completing other tasks.

There is therefore an 'opportunity cost' associated with our material resource, meaning that rational use of the money will be dictated by the other tasks we are involved in. It makes no sense to spend £100 on a new outfit for a night out if it would mean having no money left to spend on the night itself.

Skills and Knowledge

The skills that the agent possesses should also be considered in their pool of resource, as they will not have to be outsourced during the task (costing us material resource). This can include things as simple as being able to interact with other human beings but will include everything up to performing brain surgery. In any one company, there is probably a huge variety of different roles, each requiring their own skillset.

Having a diverse group of employees will increase your ability to perform a wide range of tasks. This should save you money, as going out into the market to pay for someone else to do them is going to be expensive. We could of course try to acquire these skills as part of the task we are performing, but this is likely to cost a similar amount in our time.

Any qualifications held by the members involved in the task should be a strong indication of the skills and knowledge they should possess. However, given the rate of change of information in different fields and the fact that we need to continually refresh our education throughout our careers, we shouldn't assume our friend with a mechanical engineering degree who has worked in finance for thirty years will be able to fix our car…

I find it useful to make a distinction between 'skills' and 'knowledge' when discussing types of resource. While skills are what you will use to complete the task at hand, knowledge might help to make the whole process more efficient. Knowing the steps in a process that are going to cause the most problems will help you to plan better, so you can use your skills in a more targeted way. While this knowledge is typically built up through experience, it can also be gained through education or mentoring.

As an example of the difference, my team can perform all kinds of simulations to help identify what might help to make the car faster at a particular circuit - we have a sophisticated set of metrics that we can interpret to tell us what is going to be more or less important and we use our skills to reduce this information into experiments that can be performed at the track. While this is a valid approach, it doesn't necessarily replace the knowledge we gained during the last race at this circuit. Knowing that there's one particular bump that the drivers always complain about helps you to prioritise your efforts into solutions that will make the biggest difference on the day which avoids having to start from scratch each time.

Time

Clearly all the material resource, skills and knowledge in the world are not enough to guarantee a desirable outcome if there is not enough time to complete it. This needn't only be the difference between the current time and any deadline that the task is required to be completed by; it will include additional 'man hours' that arise from having multiple individuals working on the same thing. Any parts of our process that have fixed times associated with them will need to be accounted for appropriately.

Like our material resource, we must remember that time spent doing the task in question is time that cannot be allocated to anything else, so again there is an 'opportunity cost' associated with how we spend it. The rational use of time will therefore depend on the all the tasks we need to perform at any point.

Time is somewhat special when compared to other types of resource because we find there are amounts of time that we cannot attribute a financial cost to. Imagine if you were offered £1 billion to work on a menial task for every minute you were awake for the rest of your life. Would you consider it? If you had any time to think about it, I would hope not. What's the point in having all that material resource if you have no time to spend and enjoy it? For reasons like this, we should be quite careful when considering how much our time is actually worth.

The above list probably doesn't contain any big surprises, but it should help to describe what we will be consider as resource in our task. For an illustration of how this may look in reality, I can describe some of the resources available to an F1 team.

The material resources will include all the financial resource that can arrive from either Formula 1, in the form of prize money and sponsorship deals, or injections from shareholders. Alongside this, the team will already have a significant number of assets, such as office space, a wind tunnel, various test rigs, a simulator, machine tools etc. These will be useful during the design and development of the car, but they can also be hired out to other teams or organisations for additional financial resource. There will also be several bespoke software tools, such as simulations and data processing software that are tailored to the team's analysis needs.

The skills required to participate in Formula 1 exceed the engineers involved in the development of the cars. There will be machinists producing parts, mechanics and technicians assembling them and production engineers organising the whole operation. There will also be a need for marketing skills to raise the necessary finance, and supporting operations like IT, legal and HR to keep everything moving along as smoothly as possible.

The knowledge a team has will reside among its employees and in its technical libraries, documenting analysis from races and any experiments they may have conducted. Often personnel will be brought in from other teams purely because of the knowledge that they will bring with them, as keeping up to date with how other teams are progressing will help to avoid missing out on key developments that others may be well ahead in.

Finally, the time resource available to design and develop the cars will run until the start of the first test of the season. It is possible to compete in the first race without attending this test but the competitive nature of the sport these days suggests you will struggle if you aren't ready for it. You can start your design and development process for the following year's car whenever you are finished with the current year's (or even earlier) but you'd miss out on valuable information that you could gather on the current car and feed into the development process for the next one. The total time resource will depend on how many employees you have. Certain types of resource, like wind tunnel time and on-track testing are restricted by regulation, meaning everyone gets about the same, but there are also roles that aren't tied to the yearly car development cycle at all, like those working in the supporting functions. Here, resource will probably be dictated by the projects they are participating in.

For F1 teams and any other operation, we can expect that the gain or loss of resource, the kinds described above, to influence the agent's utility. We might therefore expect the Utility Function of the agent to include elements of each kind of resource. We can imagine that if we were to lose a certain skill, like our ability to perform at a certain level at a sport, this would influence our mood and therefore overall satisfaction. For this reason, we cannot really think of the agent's resource being completely independent from their utility, in fact, the gaining of resource, be it knowledge, skills, time, financial or otherwise, might be the whole point of the task we are performing.

Uncertainty

The uncertainty in any agent's resource is perhaps not the most intuitive of all the examples we will consider, after all, projects that we encounter in our working lives will typically have a fixed operating budget, project team, timeframe etcetera, and we tend to believe we have a decent grasp of our own knowledge and skills. We can, however, consider other scenarios where there is significant uncertainty.

When filling out an online form for home contents insurance, how many of us struggle to attribute a fixed monetary value to all the possessions we would like to insure? This is likely to be a significant proportion of any individual's resource 'wealth' but the sheer quantity of items, as well as our evident inability to place values on possessions that we own (the Endowment Effect) makes estimating the total value incredibly challenging for us. Similarly, when doing the same for car insurance, we are probably aware of how much we paid initially but factors like depreciation, wear and tear and market conditions surely make it exceedingly difficult to estimate the precise value of the vehicle we're insuring.

This is where uncertainty effects are important to consider. We may not know the values discussed above precisely, but we can probably choose bounds that the true value lies between, which will be more useful than using a single value and hoping that it's somewhere near reality.

We can also consider the agent's financial resource as uncertain, as even when given a fixed budget, we can sometimes gain additional finance from investors at a later date. As an example, how many government projects are you aware of that arrive on budget? In cases where the contractors have overspent, they do not simply down tools and return to the customer to say that they have failed to achieve the project aims with the resource they had available. In this instance, they will try to secure the remaining funds they need to complete the project from the customer. Indeed, a common sales technique when bartering for contracts to make a low pitch, with the aim of securing additional funds when certain 'unforeseen' elements come to light at some point in the project. This helps to boost profit margins when the customer is tied into the contract.

When it comes to skills, in a world that is full of such a diversity of problems, just because someone has experience with similar projects is no guarantee of success in future tasks. A significant amount of learning is often required on the job, which can take up valuable project time. For example, software engineers may find themselves having to write code in languages they are not familiar with through necessity, or small changes to a systems design may have a significant impact on the processes that are required.

We can approach uncertainty in our time resource in a similar way. I'm sure we can all recall plenty of occasions where we have found ourselves rushing to meet a deadline when we had previously thought we had enough time to complete what had been expected of us.

Two factors could be at play here. Firstly, the time required to complete the task may have been estimated incorrectly. When we see the task playing out in our minds, do we consider the time taken to rectify a couple of mistakes you will make along the way? Secondly, the amount of time we *actually* have available to dedicate to the task may be different from what we expect at the start. Yes, it may be possible to complete something in a day or two with no distractions, but our lives are typically a multitude of parallel tasks that we will have to weight according to their importance. This pushes even the simplest tasks with low priority down the job list. As an example, how long does it take to physically change a lightbulb versus the time taken to actually getting around to doing it?

In experiments, people have been shown to drastically underestimate the time required to complete certain complex tasks. For an example, Kahneman showed that even when people had experience of the task they were performing, they often took twice as long as they predicted at the start of the project. This is surely something that we can all relate to. It is important to recognise that estimating the duration of certain tasks is something we can all struggle with, so your plans should be robust to this. In situations like this, it may (but probably won't) be helpful to refer to Hofstader's law, which states:

It always takes longer than you expect, even when you take into account Hofstader's law.

In instances where a task has a fixed deadline, it is worth considering how flexible this deadline is likely to be. Is it truly the point of no return, or has it just been set as a time to guide the individuals involved into how long they should be spending on it? As we have already covered (when discussing how the time dimension can affect utility), you will normally find that when a project is progressing well and giving good returns, arbitrary deadlines set at the start of the project will sail by without a second thought.

When certain types of resource have opportunity costs associated with them, we should consider that these are uncertain too. Do we really understand how long the alternative uses of time will take, or the costs involved in these other options? While we expect to pay a fixed amount of money to have a crack in a wall fixed, there is unfortunately a chance that this crack could point towards a more serious problem with the foundations that ends up costing many times more. These must enter into our decisions when considering the price we will pay, or expect to receive, to complete the task we are focussing on.

There may be other reasons why we cannot use fixed values for resource parameters, but this should hopefully convey the need to consider the uncertainty in our resource when modelling a particular task.

10

A Complete State

If you think back to the person you were several years ago, I imagine you will notice some differences. There may have been some music or a TV series that you were obsessed with then that you aren't so passionate about now. I suspect your bank account looks slightly different, hopefully in a good way. You have probably learned things about the world that you didn't know back then and had a whole range of new experiences. The life of an agent is one with many twists and turns and it makes no sense to think of the one in our model as a static, inanimate entity.

In order to allow our agent to evolve over time, we should consider the concept of 'state', defined here as the minimum amount of information we need to define the Utility Function and resource quantities in sufficient detail for the task in question at the current point in time. This state should include any information that is relevant for the rest of the task. Some points you could consider are:

- Utility Function
- Reference point for utility calculations
- Financial resource remaining
- Skills and knowledge
- Time resource remaining in the project

We may add or remove items from the above list depending on the aspects of the agents that we need to track during our task. This information will be important for making decisions in 'time-marching' scenarios, where Utility Functions and resource amounts could be changing frequently.

For our agent to be capable of evolving over time, it must be able to transfer information to other areas of the model and receive information back. The targets our agent has will be passed to the control process, where actions will be performed based on these and the measurements taken from the environment. Similarly, our resource will pass out from the agent to be used in other areas of the process. We might also receive some resource back after the control process has been performed and we've interpreted our response from the environment, and furthermore, receive utility from our inferences on our measurements and any resource gains and losses. This utility may alter the shape of the Utility Function and the reference point for future decisions.

We cannot ignore the fact that these information transfers will also have uncertainty associated with them. If information is coming in from all different corners of the model, it's possible that we could lose something in the transfer. Take the example of our target - I'm sure you can think of examples where you've requested something of someone only for them to go away and perform something different to what you had requested, maybe a tradesperson who misunderstood where you wanted something installing? Or perhaps at work a task was given to an inexperienced employee? We may have a clear idea of what we hope to achieve from the task in our own minds but describing this to others is a skill in itself. At the end of the occasion you're thinking of, I'm sure the person in question was as surprised as you were to discover what they had done is not what you wanted.

We can apply this thinking to the resource transfer as well. When we set our expectations for how much time we are prepared for someone to spend on a given task, can we be sure that the message is understood, and that everyone involved knows which skills they are supposed to apply to the problem? Is the financial resource available obvious or is there a risk an eager team member could spend all the budget on day one? We should ask ourselves these questions when passing information between the individuals involved, as checks like this will reduce the uncertainty in our process and should ensure a better result. Any schoolteacher will tell you that even with the clearest explanations in the world, you can be amazed at the range of interpretations that can exist. It's important to set tests every now and again to ensure the messages are getting through.

On the other side of the problem, we might fail to interpret the results coming in from different areas of the model correctly. When financial resource comes in from our control process, maybe we misunderstand how much can be applied straight back into the process, particularly if someone else has already allocated it to paying off other costs. It may be that we fail to interpret how the results being fed into us should affect our utility, or a series of unintelligible figures on a financial statement may have absolutely no meaning to us. It will also be difficult to understand precisely how much time all the individuals involved have spent on the task you are monitoring. Again, all these uncertainties will make it difficult to infer the correct outcome for your task and anything we can do to reduce the level of uncertainty will probably lead to a better outcome. Even if we realise that things haven't gone as well as we might hope, this is surely better than putting our faith in something that isn't correct?

If we are performing the entire processes ourselves, the problem is much simpler, since communicating things to ourselves should be straight forward enough. If not, I'm sure you'll be able to find plenty of self-help books that can help you be 'true to yourself'. In other situations, monitoring the state of the agent in the model is going to take some effort, but this will certainly pay off when a far better outcome from our task is achieved.

11

Agent Summary

In this section, we have described the most important characteristics of the agent that we will use in our task model. We've described how the agent consists of a Utility Function and their resource, and that when playing the task out over time, this information is held in the agent's state at any fixed point.

One of the key ideas we've introduced is the concept of utility, which is invaluable as it allows us to compare the preferences of the agent using a single metric. The agent's Utility Function should be able to of convert the outcome of our task into a value of utility. This function's inputs will be the elements of the outcome that the agent considers important, which will not only be a financial outcome but also more qualitative ideas, like the satisfaction of those affected by the task. The final function will be subject to significant uncertainty, not only because the methods we use to build it will be imperfect but also because the agent's preferences are likely to change over time, as well as being incomplete (it is unreasonable to expect the agent to be able to assign a utility score to every possible outcome ahead of time).

We can make some assumptions about the appearance of this function when considering how human beings will normally behave in the face of decisions. The agent's utility for money will be s-shaped with a discontinuity at zero that represents a human being's tendency for loss aversion. We expect that the gradient of the function will reduce as the sums of money increase, which can be described as the 'diminishing marginal utility' of money. When building the remainder of the Utility Function, we can use indifference curves to help us express how the agent considers commodities trade with one another.

We can use this Utility Function when making decisions about how our task progresses. In these decisions we can think of maximising the expected utility in every decision the agent makes as rational behaviour, however, the science of Behavioural Economics teaches us that we cannot necessarily treat human beings as rational in every decision they make. Prospect Theory describes how human beings tend not to treat very high or low probabilities in the same way as a completely rational agent, and that each decision relies on a reference point of the agent - a function of the agent's experiences up to that moment in time. Phenomena such as anchoring will make predicting the behaviour of different individuals more difficult, as they will rely on knowledge of the history of decisions up to that point.

When we consider the agent's resource, we are not simply considering the money available to complete a specific task, but also other important quantities like skills, knowledge and time. We must also accept that these quantities have significant uncertainty when making decisions on how to use them.

Finally, we have introduced the concept of state that will define the agent's condition at each point in time. We expect this state to evolve over the course of the task, influencing the resource and Utility Function.

The appendix of this section will describe some simple methods for building our Utility Function, which will hopefully help you get started on some practical implementations. Now we have the agent of our task, we need to understand the situation they are operating in. We will call this the 'Environment'.

Part One Appendix

Here's One I Made Earlier

Here we will discuss the creation of simple Utility Functions to represent the agent of a task. We will consider the following cases:

- Constructing a Utility Function for a single individual
- Constructing a Utility Function for a group

The first case, in which we are only interested in one person, is simplest. We are trying to establish how much they value sums of money relative to other amounts.

We will split the task into two parts:

1. Build the 'gains' half of the utility curve
2. Build the 'losses' half of the utility curve

For step one, we should first understand the amounts we are dealing with when we talk about gains and losses for this task. If we are going to be making trades with sums of money of the order £100, it would be senseless to evaluate our Utility Function over the scale of millions. Similarly, we shouldn't be asking questions about wagers of the order 10 pence.

Let's say we have settled on magnitudes close to £10. Now we can start to use the following type of wager to help build the Utility Function:

50% chance of receiving x and a 50% chance of receiving nothing, or y for certain

A question we can ask is the lowest value of 'y' the individual in question would take for a given value of 'x', for example:

50% chance of receiving £10 and a 50% chance of receiving nothing, or y for certain.

If we were to look at the above wager, let's say the lowest value of 'y' an imaginary person would take is around £4. We can interpret this as the utility of a 50% chance of receiving £10 being the same as the utility of a 100% chance of receiving £4, or the utility of £10 is double the utility of £4.

Now we can move to slightly different kind of question:

50% chance of receiving x and a 50% chance of receiving nothing, or £10 for certain

We can ask, what is the lowest value of 'x' that would make this wager worthwhile. Imagine this is around £25 for our imaginary participant. Using the logic we used previously, we can say that the utility for winning £25 is double the utility for receiving £10. We now have four points that we can plot on our curve:

1. Utility of £0 is zero
2. Utility of a £10 gain is defined as 1
3. Utility of a £4 gain is 0.5
4. Utility of a £25 gain is 2

We can now plot these on a graph and draw a curve through them. This is shown in Figure 10. From here we can choose to stop or carry on investigating larger or smaller numbers as we please, perhaps varying the odds we are using. We should be aware that this method is very sensitive to inaccuracies in the values already given, as we are anchoring to utility values at zero and £10 and building up from there. Human beings tend to round to numbers that seem neat (you're unlikely to get responses like £4.3283435....) so it's worth repeating the procedure a few times to ensure consistency of shape (choosing different values for the variable parameters each time).

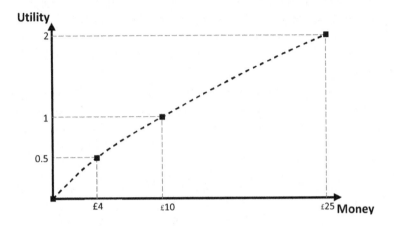

Figure 10. Gains half of utility curve

We have already demonstrated in our discussion of Prospect Theory that we should stay away from using very low or very high probabilities in these wagers. Human beings tend to misinterpret probabilities close to the extremes, so it's best to stick to close to 50%.

Alternative procedures use different 'standard forms' of utility curve. A quick internet search of 'Hyperbolic absolute risk aversion' will give the interested reader more information on this standard form.

Now we have the 'gains' side of the Utility Function, we can focus on the 'losses' side. Here the procedure is very similar, except now instead of dealing only in gains, we consider losses as well. The wagers can take the form:

50% chance of receiving x and a 50% chance of losing y

We would ask the participant for highest value of 'y' where they would be prepared to accept this wager, given a value of 'x'. By using the values of utility we have already collected, it should be trivial to now populate the losses side of the plot. For example, if the value of 'y' in the wager above is £3 when 'x' is £10, we know that the utility of a £3 loss is '-1' for this individual. This is because the value of utility we attributed a gain of £10 was '1' in the previous step. By performing this wager using the values of utility we have already collected, we can construct the rest of the curve. The figure below shows a completed example.

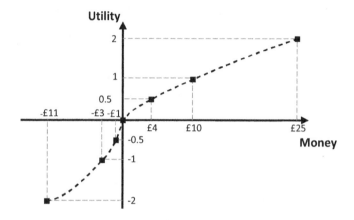

Figure 11. Completed utility curve

Note that the magnitudes for the losses would be expected to be greater than for the equivalent value as a gain. This is due to a human being's loss aversion.

Let's move on to our second case, where we're dealing with a group of participants. The issue with using the same procedure is we will naturally get different answers from each person. What we are left with now is a weightings problem to define the final version of the Utility Function that we are going to use. There are no hard and fast rules that I'm aware of that cover how this could be done, but we can consider a few alternatives:

- Equal weightings for each person - good for groups where everyone has an equal stake in the agent
- Weightings based on each individual's stake holding in agent - useful for groups that do not have equal shares in the agent

- Taking the median curve value at each point - good for removing outliers, such as individuals who are risk-seeking or extremely risk averse
- Voting on preferred shape - again, good for removing outliers

Once we have our preferred weightings scheme, we can apply the relevant weights to each of the individual curves to get the final version that we will use in our model.

Constructing indifference curves is somewhat simpler as we can go back to working with familiar metrics and units. All we are looking for now is how much of a commodity our agent would trade with quantities of the others being considered. Clearly with more commodities to trade, this task gets more difficult. We could assume simple relationships between commodities, but we will struggle to capture non-linear effects that occur for strange combinations of each.

We can create fixed-interval utility contours on our indifference curve plot by referring to our utility for money curve that we generated here. For each pair of commodities, we can just create a new indifference curve based on having different amounts of money that correspond to fixed utility steps. If we wanted to create indifference curves for gin and tonic water for fixed utility intervals, given the Utility Function we created above, we could ask how much of each commodity they would demand if they had £4 (0.5 utils), £10 (1 util) and £25 (2 utils). Cleary, we would need to interpolate our map to get something for 1.5 utils.

Part Two

Environment

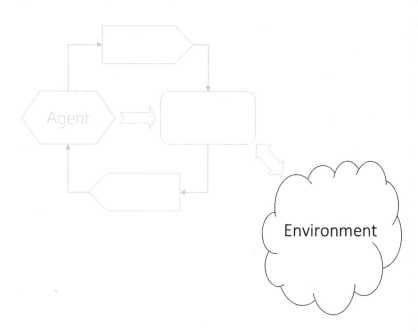

12

It's Not Just About Having the Fastest Car

In 2010 I was working with the Williams F1 team. The racing season that year had been reasonable but not exactly one that would live long in the memory, had it not been for one remarkable qualifying session in the penultimate race of the season...

The Brazilian Grand Prix was supposed to be a battle between the five main championship contenders. Williams was expecting to be battling in the mid-field, with other teams that couldn't mount a serious attempt at the championship. The attention at the factory was firmly set on the following season, where there was a lot more up for grabs.

The rain began to fall on Saturday morning prior to the final practice session. This was always something that mixed things up and the team was optimistic that they could make the best of it to gain a better result than they would have expected in the dry. Qualifying was split into three parts, with the aim was to make it through to the final session by avoiding the 'drop zone' (the seven slowest cars) in the two previous sessions – only ten cars would make it through. In the season so far, Williams had found it difficult to get into this final session at most tracks, without mistakes from the drivers in faster cars. Sometimes other teams have just done a better job and you're left to pick up the scraps of what's left.

It's Not Just About Having the Fastest Car

The drivers for that season were Rubens Barichello, a veteran of championship winning teams towards the end of his career, and Nico Hulkenberg, a promising rookie in his debut season. Both drove brilliantly in the wet to make it through to the final session, something that already constituted a better-than-average result for that year. It was what happened next that took everyone by surprise.

The rain began to ease during the third session. This is one of those occasions where the drivers and the engineers can gain a huge advantage by being on the right tyres at the right time; switch to dry tyres while the track is too wet and they will lose all temperature and provide no grip, but stay on wets too long, and they will overheat and you're likely to end up slowest of the runners.

The first drivers to switch to dry tyres found the going very difficult, with some spins and slow lap times. The Williams drivers left it until the very end of the session to switch, which was risky as it would probably only mean one or two laps on these tyres. There would be no slow build up - it was essential to deliver the best time straight away.

Nico surprised everyone by going fastest on his first attempt on the dry tyre, but there was still plenty of time for others to respond. Rubens was fast as well but a little further behind. Amazingly, the other drivers were struggling to improve, with few able to challenge Nico's time. It was then that Nico put in the lap of his life, beating his own time to lead the field by a whole second! It was unbelievable, gaps that large were rarely seen between cars in qualifying, let alone at the front and by a rookie driver, driving at that track for the first time in an average car.

That's how the positions stayed. Rubens managed a very strong sixth, in what constituted by far the best qualifying session for the team in many years. There was obviously much to celebrate. It was yet another lesson in keeping your head and focusing on your own performance to get the job done. The car on pole position wasn't the fastest, and nor did it have the best driver; it was the one that made the most of the conditions in the *environment*.

13

Learn the System

From where you are sitting now, how would you describe your environment? If you're at home, you probably have electric lights that you can turn on if it gets too dark to read. You may have a central heating or air conditioning system that is keeping the room close to your desired temperature, and maybe there are other people around who are demanding your attention. If you're outside, you'll be experiencing whatever the weather is like today. You may have your phone with you that is giving you notifications of people wanting to speak to you, or feeding you information about something that's going on elsewhere in the world. Wherever you are you will be surrounded by 'systems', either natural or manmade. These are affecting you in many ways, and you, in turn, will have the ability to affect some of them, but not all.

These systems, without your input, will behave as they always behave; the heating will carry on working without you having to give it any attention, the weather will be sunny on some days and raining on others and other people will be going about their daily lives. However, with your input, you can control your own temperature, be it by changing the air temperature or simply your clothing. You can communicate with other people to ask them a question or tell them some news, you can even request work from people or objects to help you achieve whatever you have your heart set on doing at that time.

We will define a system as a process consisting of multiple elements that behaves according to its own rules. It should be possible to draw boundaries around each system and separate the way we interact with it into inputs and outputs. These are the building blocks of our environment and the concepts that we must understand before we can make progress with our tasks.

Systems can be categorised in many ways. Some examples of different systems you might encounter are listed below, most of which may be very familiar and no doubt you'll be able to come up with many more yourself:

- Natural
    - Biological systems
    - Weather systems
    - Ecological systems
- Man-made (designed)
    - Financial systems
    - Cultural systems
    - Engineering systems
        - Electronic systems
        - Mechanical systems
        - Hydraulic systems
        - Pneumatic systems
    - Information systems
    - Political systems
    - Legal systems

You should be able to find more literature on each of these examples than you could read in a lifetime. While we will cover common aspects of different systems and the environment across all tasks, we will not be discussing specifics about how to model systems from different subject areas unless they are particularly relevant. If you are an engineer, you will probably have a good grasp of the physical laws required for engineering that you use frequently. If you are an economist or working in finance, you will probably have an appreciation for the behaviour of markets and are able to create effective models for your purposes. What we are more interested in here are the universal behaviour of systems and what these can tell us about how we should approach the task in hand.

When it comes to our modelling task, the environment will be the stage on which the task is played out, including all individuals and processes that could be encountered, governed by scientific and empirical laws that will need to be accounted for. In this section, we will discuss how the environment can be manipulated and how it affects us, along with the processes that are being carried out and how they behave. The aim is to understand what we are dealing with so we can choose our behaviours to suit the outcome best.

In my daily life, the systems I experience are linked to the environment around Formula 1. These are incredibly diverse. Of course, the engineering systems within each of the cars will have a big part to play, as will the systems that define the condition of the driver, both mentally and physically. The weather on race day is very important and starting a race on the wrong tyre for the conditions is a very reliable way of moving yourself to the back of the field very quickly. There are also financial systems at work; teams with the best cash flow have a distinct advantage when it comes to developing the cars and signing the best drivers and personnel. Don't underestimate the amount of politics at play either; should your team get behind the latest proposals for regulation changes or could they harm your position? The best teams on the grid will navigate themselves through all these systems and hurdles to put themselves in a championship winning position. Our ability to make predictions about how things will play out in these systems is therefore of great importance.

One of the key things we are seeking when it comes to the modelling of these systems is how to predict the future. We aren't really very interested in what has already happened (other than the information it has provided us with to tackle problems that we might one day need to contend with, be it financial resource or learning of how a system might behave from this point forwards); our view is that the future is ours to turn into whatever we want, and being able to influence it by understanding the behaviour of the systems we will meet is vital.

Content of This Section

In our discussions, we will be covering the ways in which we can interact with our environment. These are the inputs and outputs that we have to guide the progress of each task. We will show how these have uncertainties associated with them that will make the reality of the actual state of the environment unknowable.

Next, we will discuss the behaviour of the systems contained in the environment and their structure, which leads onto the subjects of Complexity and Chaos. These characteristics will be common to many of the systems we come across as part of our task, so understanding their effects will teach us how to deal with them most effectively. Systems that can adapt to changes in other areas of the system can exhibit some very strange behaviour, which we will cover under the heading of Game Theory.

We will find that the systems in our environment impose certain constraints on how our tasks unfold, and we will discuss different types that we may encounter and whether there might be anything we can do to avoid them when necessary.

With all this, we should be in a better position to predict how our environment will develop into the future and therefore how we should behave to get the most from it.

14

Garbage In, Garbage Out

Let's start our discussion with how we can interact with the environment. Like any system, our environment will have a series of inputs and outputs that we become familiar with to achieve our objectives. Inputs are the actions that we make on the environment, while outputs are what we get back in return. While outputs can come about because we have made inputs into the system, there will also be times that we receive outputs that have nothing to do with our actions up to that point in time. We will refer to these as 'disturbances'. Before we can think about how the inputs and outputs of our environment might look, we should consider the 'extent' of the environment we are in, which is probably broader than you imagine.

What do you have absolute authority over? Anyone with children will probably agree that you have extremely limited authority over any other human beings. Everyone has a mind of their own and will only do as you say if they believe it aligns with their interests. Similarly, anyone who has their own PC will understand that sometimes things do not go as you'd intended. Finally, anyone who has ever dropped a glass onto a hard floor will be able to testify that maybe they do not even have complete control over their own movements.

Using this definition, we can think of our environment as beginning at the boundary of our own thoughts containing everything else, as there is nothing over which we have absolute authority. We're certainly not powerless to have an influence on the environment, we must just be mindful that it will come with a degree of uncertainty.

The extent of the environment that we want to model, however, will depend on the task at hand. If we are talking about a complex industrial project with various stages of design, manufacturing and distribution, we probably don't need to consider the control we have over our own muscles. However, if the task is one of moving a delicate family heirloom from its current position into storage somewhere else in our houses, we probably should plan for a bit more time than it takes to move pillows.

The discussion of the environment's inputs and outputs is probably best centered around a familiar example. For this I've chosen the task of driving to work from home. This is an example that we'll revisit frequently over the course of this book, as it has many nice features. It has is a very clear objective, that of getting to work in the shortest possible time with a minimum amount of stress. The environment that it takes place in is very complicated, with many other road users, traffic laws and possible disturbances. The process of driving a car is also incredibly complex, with a great deal of training and practice required (it's a wonder that any of us get anywhere at all!). As well as these features, this is a task that many of us will spend an awful lot of time doing and is therefore something that can significantly affect our moods – often for the worse. Getting the most out of this task will be very important for a lot of people.

The systems we are considering can be reduced to the car itself and the road system, with its various rules and other road users. The agent is us, the human being driving the car.

The inputs that the agent can make to the environment are defined by the capability of the agent and the systems they are interacting with. As somebody with a driving license, we should be capable of driving the car. We can press the accelerator pedal and the brake, change gear and move the steering wheel, along with a few other controls such as the indicators, headlights, wipers etc. These inputs should give us reasonable control over the car, but this authority is not infinite, as they are constrained by the capabilities of both ourselves and of the car. We will not, for example, be able to bring the car to an immediate stop if we see someone crossing the road, nor will we be able to exceed the maximum speed of the car.

The car itself also makes inputs to the road network system. From the perspective of others, it can change lanes on a dual carriage way, it can emerge from a side-road onto a main road, or it can join or leave a roundabout. As a driver, we will be able to choose our route as an input, including the direction we will be taking, roads we will use, the speed we will be going, the time we will be waiting at junctions. Our car will affect the other road users we encounter, for instance, any queues will be one extra car in length, any one we pull in front of may need to adjust their speed or direction and anyone behind who wants to go faster than us will have to overtake.

Outputs are defined as the feedback we get from the environment back to us. In our driving illustration these are the things experienced by the driver, which could be the colour of a traffic light, the presence of a car in front of us, or a change in the gradient of the road. We've already discussed how some of the outputs can be a function of the inputs we've made to the environment, such as the response of the car as we turn the steering wheel. Other outputs will have nothing to do with our previous inputs. This could be a person stepping out onto the road in front of us without looking; these are the disturbances to our progress.

Something we will ignore at this stage is the crucial process of measurement. This is the step of interpreting these outputs, so we can decide how we're going to act. The problem with measurements is that the step of interpretation is open to errors, meaning we cannot think of them as perfectly accurate reflections of reality; the outputs of the environment can be regarded as the 'truth' but the measurements will only be an approximation.

Uncertainty will again arise in both inputs and outputs. We will find the outputs of the environment difficult to predict, given the complexity of the systems involved, but we also cannot guarantee that the inputs we make are exactly as we intended either. When we press the accelerator or brake pedal, we are not choosing absolute rates of acceleration or braking, as if we were putting numbers into a computer program; we are choosing whether we want to change our acceleration a little, with small movements of the pedal, or a lot, with large movements. We can sometimes be surprised by the response of the car to our inputs, which may be particularly true if just starting to learn or if driving a new car for the first time. Anyone who has ever tried pressing the brake pedal with their left foot for the first time, particularly if it's more familiar with pressing the clutch pedal, is probably very aware of the imprecision a small change in style creates.

One of our children's favourite toys is a so-called 'hand-steady game'. There are various kinds of these games, but they usually require the player to move an object, normally a hoop made of conducting material, around a piece of wire. Touching the wire will complete a circuit and cause a buzzer to go off to indicate you've lost the game. It's a very simple idea, one where you are only competing against your own imprecision. Someone with complete control over their movements will not really understand the point of this game as they should be able to complete it very quickly indeed. Needless to say, neither the children nor their parents fall into this category. When we consider our inputs to the environment, we should think of it as if we are playing one of these games. That's not to say the imprecision will lead to significant issues, but that doesn't mean it isn't there.

In my industry, input uncertainty appears all the time in the process of manufacturing components that fit to the cars. Design engineers will not set precise dimensions on the manufacturing drawings they produce, as machining to high precision will take a great deal of expense and time. Instead, they will set 'tolerances' on each measurement. If the dimension on the manufactured part fits between these bounds, it will be considered acceptable. If these bounds are set too loose, parts will be cheap to manufacture but risk not fitting together when assembled, whereas should the bounds be too tight, things get expensive and start to take a long time. Either way, the final dimensions of the part cannot be known with certainty beyond what is input to the process, i.e. the manufacturing tolerance.

The uncertainty in the outputs will be dictated by our ability to model the entire system. We can model the overall journey time as a function of the route, the time we are travelling (journeys in the middle of the night will be faster than those in rush hour) and the measured state of the traffic. This is something that satellite navigation programs can do reasonably well; some may even predictively account for how situations are likely to change in the future. However, do not expect the time predicted by the program to be precisely the time it will take you to reach your destination, even if you drive in exactly the manner that's expected. Uncertainty arises because conditions are liable to change all the time. People in front of you could choose to slow down to let other people out of side roads, level crossings can close just ahead of you, you may be unable to overtake a cyclist or tractor for several miles. These are all sources of noise in this system that will undoubtedly affect the final journey time.

We will see that the same characteristics arise in other environments, with completely different systems, and completely different inputs and outputs. Take the example of organising a party for friends and family, an environment that consists almost exclusively of other people. Our inputs and outputs are the conversations and communications between everyone involved; we can ask questions, like whether they can attend, we can give instructions, like asking them to pick up some of the food from a local shop, or we can ask for information, like whether they would like to bring a guest or if they're coming on their own.

The systems we're dealing with are the guests' daily lives and whether they can accommodate the party invitation into their schedules. If one of your prospective guests has a diary entry they have already committed to, it's unlikely they'll attend. If someone you would normally ask for help in gathering supplies isn't able to drive at the time, they probably won't be able to pick up the food you need from the shop. If you have neighbours with young children, they will probably prefer if there wasn't any noise coming from your house in the middle of the night.

Of course, there is uncertainty in all these systems. Before you invite anyone, you won't know whether they are able to attend, and it would be senseless to buy all the food before you know roughly how many guests to expect. Even if they've confirmed, it's not impossible that people could drop out. People who originally declined could decide they are able to attend after all and others could ask to bring additional guests after sending the RSVP. People's lives are complicated, so it's important not to take their responses as certain. It might be that your neighbours have decided to spend that weekend away and you can make a little more noise, but if you've misunderstood this message or got the date wrong, you're likely to end up with a problem.

Something that we've touched on throughout these descriptions is the idea that the systems we're dealing with are likely to be complicated, with many parts that can make our modelling difficult. In the next chapter, we can start to see how mathematical ideas can help us to understand these systems a little better and, most importantly, teach us the limitations of our models and why we need to be careful when putting them to use. We'll start with the concept of 'Complexity'.

15

Beyond Merely Complicated

All Formula 1 teams observe a two-week mandatory shutdown at the beginning of August. This is inconvenient if you don't have children and would prefer to avoid paying school holiday rates for your summer vacation, but it does mean you get a full two weeks of uninterrupted break from the stresses of the job. Most years, I have spent the shutdown in Cornwall, enjoying the scenery and spending time with family. This part of the world has some excellent surfing beaches, and the waves draw some of the country's very best surfers for these summer months (my ability is definitely not in the same league).

If you stand at the top of the cliffs behind of one of these beaches and look out to sea, you'll see what looks like a perfectly flat ocean. It's hard to believe that the waves that end up crashing on the shore have all originated out there. Closer to the shore is where the waves become more visible. These typically rise out of the surface in nice, ordered rows before breaking as the depth of the water can no longer sustain their size.

When you're in the waves, they don't look quite so serene. Ripples on the surface interfere with the height of the water, causing the waves to break earlier in some areas than others. After the waves have broken, you get the turmoil of the white water, which can knock you off your feet as it churns up all the sand from the beach. Finally, at the smallest scales, you can see small local micro-currents in the water, with small eddies and areas of turbulence. The activity that is going on at all scales is mind-boggling to behold. This is truly a Complex System.

The rules that govern how water molecules interact with each other are reasonably straight forward, however, these simple interactions build into features with huge complexity when grouped together in huge volumes which are exposed to the geographical systems of the Earth. The path of any single drop of water is impossible to predict in this environment, though with a big enough boat, we can more or less disregard the workings of the system at these scales and treat the water as basically flat, with a series of currents that can help you get to where you need to go.

We will undoubtedly come across systems that are similarly 'complex' during the tasks we encounter on a daily basis. This term covers more than just the merely complicated; specifically, we will use it to describe systems where the parts interact with each other in a variety of ways, usually with a reasonably simple set of rules. This will give rise to 'emergence' of 'higher order' characteristics.

Take another example, that of a pneumatic tyre. In a tyre, grip is generated by the interaction between polymer compounds in the rubber, both with the road surface and between themselves. Molecules transfer kinetic energy between themselves, heat up and change properties further. After prolonged, heavy use, the tyre will behave differently to when it was new. Despite the huge complexity of the system at the molecular level of the hierarchy, tyre properties are reasonably predictable, and we can create models for how they behave without having to worry too much about the interactions at the very small scales; we simply study the emergent properties, such as 'grip'.

Complex Systems will usually have behaviours that systems that are merely 'complicated' will not. Emergence is key here. The gearbox of a Formula 1 car is undoubtedly complicated with many different hydraulic, electronic and mechanical parts, but it is lacking in any emergent properties. It is designed to convert the power at its input into torque and speed at its output and this is what it achieves (if done successfully). There is no equivalent of turbulence or grip in a gearbox.

Even when discounting emergence, we should be able to identify a Complex System by other behaviours that most will exhibit. Firstly, these systems will have a tendency for self-organisation, meaning they will organise themselves into some kind of structure without any interference from outside. Biological systems that are reached through a process of evolution could be considered an example of this, much as creationists would prefer to deny it. The idea of civilisation emerged in early humans due to the interactions occurring between individuals who saw the benefit of grouping together to share skills and effort. The organisation of individuals into countries, cultures, companies and so on does not appear in simpler systems. Similar behaviour can be seen in schools of fish or flocks of birds. The rules that dictate how fish and birds travel together are simple, but they can lead to spectacular displays in nature.

We can also expect the frequency of extreme events to be higher than probability laws, like the 'normal distribution', would suggest. This behaviour can be described as having 'excess kurtosis' or being 'heavy-tailed'. Examples could be booms and crashes in stock markets, periods of extreme temperature in weather systems or population booms or extinction events in plant and animal species. These extreme events occur due to the interactions that are happening in the system. Re-organisation can lead to extreme positive or negative feedback that can force the system in one direction or the other.

Finally, Complex Systems can exhibit 'chaotic' behaviour, where motion at the smallest scale can create changes in the emergent properties higher up the hierarchy. We do not require a Complex System to begin to see chaotic behaviour; this subject stands alone and understanding its influence will be the focus of the next chapter.

Respect the Hierarchy

We should be able to describe a Complex System in the form of a hierarchy, with the components at the smallest scales occurring at the bottom and the highest order emergent properties appearing at the top. The hierarchy of the system is something we must pay attention to when making decisions about modelling; we should see that each level is governed by its own set of laws that are largely independent of other levels.

If we go back to our ocean water example, at the atomic level the behaviour is governed by the laws of physics, with fundamental forces holding individual particles together in atoms. If we go a level higher, the water molecules themselves follow the laws of chemistry, with atoms forming bonds with one another. There may also be minerals dissolved within the water that affect its properties. Go another level above this and we have laws that govern the fluid flow, including conservation of mass and momentum. We can also start to see areas of turbulence, with eddies at different scales all interacting with each other in ways that defy explanation at smaller scales. Finally, at what we will think of as the top level of the system, we have the geographical features of the waves, the weather and the tides, which are governed by various systems with scales that can be as large as the planet itself. Each of these layers can interact with those around them as energy can cascade up and down the hierarchy, affecting behaviour at all scales. Thinking of the hierarchy in this way will be very useful to us when we set out to create our model.

Let's take another example of the economy of a country, or indeed the whole world. This too has a hierarchy that should be familiar to us. At the bottom, we have individual human beings and the machine tools they use (arguably we could go down further to the individual machine components and human organs/cells/atoms but life's too short). A level above this are the teams and departments that they group into, with the companies that are created from them occurring next. On the next layer are the individual markets and finally, the emergent properties of the economy sit at the very top. This could be the nation's wealth, well-being and outreach. See the diagram in Figure 12.

In this hierarchy, the behaviour of each element at each level of the hierarchy will affect the levels around it, for instance, the performance of individual members of a team will affect the performance of the team as a whole. Similarly, the management of the teams will affect how each person below them in the hierarchy perform, whilst also affecting the performance of whole company above them. It's with links like this that you can see how behaviour at the very bottom of the hierarchy can find its way to affecting the emergent properties of the very highest orders.

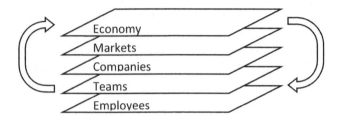

Figure 12. The hierarchy in the economy system

Take the example of the stock market 'Flash Crash' of 6th May 2010, where a trillion dollars was wiped off the value of the US stock exchange for around half an hour. This calamity was remarkably blamed on a single, British trader, Navinder Singh Sarso. The precise mechanisms that were used to trigger this crash are quite complicated, but they involved placing a series of large wagers, amounting to hundreds of millions of dollars, on the crashing of the stock market. The plan was to cancel these before they were actioned, where they would need to be paid for. The very presence of these wagers was enough to trigger the crash. More remarkably, Navinder was not working for a large trading firm or indeed for any trading firm at all. All of this was masterminded from his parents' spare bedroom.

If we want to capture the behaviour at a particular level of the hierarchy, we can create the model at this level, without worrying about detailed modelling of the levels above or below. For example, if we want to know what the state of the economy will look like in a few years, we can model how we expect the markets will evolve, without worrying too much about the individual companies or their component parts. Similarly, if we are interested in how we should budget for changes in our company to receive the best return on investment, we can study how individual departments behave, without modelling interactions between individual employees. This simplifies the problem of modelling so that we can create something useful without having to worry about infinite complexity.

Over the course of the above examples, we have covered both major types of Complex System. These are:

1. Complex Physical Systems
2. Complex Adaptive Systems

The tyre and ocean systems above are examples of complex physical systems, as the rules under which they operate are all contained within physical laws, e.g. laws of thermodynamics, laws of motion etc. An adaptive system can change its behaviour when exposed to certain actions; a country's economy is a good example of an adaptive system. This can change its behaviour as commodities increase or reduce in value or as the population becomes more, or less, skilled in certain ways. Elements of physical systems will continue to obey the same physical laws, regardless of the state of the rest of the system. This adaptive behaviour creates another element to our discussions.

When the environment has multiple, intelligent agents all acting at the same time, i.e. a Complex Adaptive System, we need to consider two ways that agents can behave towards us (the agent we are modelling). They can be cooperative or competitive, which leads us to an intriguing subject that describes how agents can interact and how this can affect their environments.

Playing Games

'Game theory' is a subset of the broader subject of 'Decision Theory', which we touched on in the previous section. Game theory covers decisions made in the presence of other agents, where all agents have influence on how a system evolves. When we add extra agents to the equation, some very interesting things can happen.

In our environment, we are likely to meet other agents. These can have different attitudes towards us, and we can have different attitudes towards them. If they believe that working together will lead to a better outcome for themselves, they will be inclined to cooperate with us. Hopefully you don't have to look to far for examples of this in your own life; our friends and families will normally support us in our lives when we need help, in the same way that we would for them. When we visit shops to buy groceries, we are engaging in cooperative behaviour; we benefit from what they can sell and they benefit from our money. Even agents on larger scales, like companies and countries can form alliances when they think it's in their best interest. 'Platform Sharing' describes manufacturers sharing the basic architecture of their cars to reduce design and manufacturing costs and is reasonably common in the commercial automotive sector. Similarly, countries across the world can form alliances when they believe they can both benefit, maybe in the face of competition from a larger economy.

Of course, you will also meet the opposite kind of agent in your lives: your competitors. Some people will have the attitude that taking utility from you is an efficient way of increasing their own; in many ways, the basis for a capitalist society. With a lot of competing companies in a market, those involved are encouraged to push themselves harder to provide better value for money than the competitors can offer, which ensures their continued survival whilst also leading to innovation for the customer. Any member of this market who finds themselves unable to compete will be pushed out - the basic premise for Evolution by Natural Selection. The species that are best adapted to their environment will be able to survive in greater numbers and spread their genes further. Of course, we needn't think on such grand scales when talking about your competitors in day-to-day life. When a vacancy at the level above your current role opens up, you may want to go for it, and hope to rise above the competition.

Another interesting position to look out for in adaptive systems is that agents can be both competitors and in cooperation simultaneously. Maybe you have friends that play on a different team in a local football competition. You may well find yourself going in for a tough challenge with someone, while at the same time expecting them to look after your kids at the weekend. On a larger scale, Apple and Samsung were involved in several lawsuits throughout the 2010s as each believed the other's phone technology infringed on their intellectual property. You may or may not be surprised to hear that at the time Apple was the biggest customer of Samsung electronics.

Game theory explains that when you interact with either cooperators or competitors, you are engaging in 'games'. This term seems to somewhat trivialise your interaction but in reality, this term just describes a situation where there is a limited amount of utility available to both of you, and you can choose to either share it, or allow one party to take more than the other. Consider cutting a cake at a birthday party. The person with the knife can decide on how much everyone receives, and a perfectly valid solution is to take the whole thing for themselves and leave others with nothing. This would leave them rich in cake, but light in friends. In this scenario, you will probably find most people will divide it into equally sized pieces, while the extremely devious give themselves slightly more, but not to the extent that they can really be accused of manipulating the situation for their own advantage.

Let's first consider how cooperating parties tend to behave. Cooperation is a very desirable thing for the progress of humanity; being able to pool resources together, rather than waste energy competing must surely be a more efficient way. However, there is a big problem. Cooperation is normally an 'unstable equilibrium' (more completely, an unstable 'Nash' equilibrium, after John Nash who first worked on the concept). That is, in a series of linked decisions where the agents have been cooperating, any party that deviates by taking an opportunity to throw the other under the bus (to make a gain for themselves) will typically spell the end of the cooperation. This is reasonably intuitive. If you think of relationships you've had in the past, it's likely you will agree that trust takes a great deal of time to build but only an instant to shatter. The party that betrays the other will probably be wise to avoid seeking the trust of the betrayed party for a long time to come.

As an aside, something we can do to help here is to make the cost of breaking the cooperation very severe. Companies that choose to cooperate can have systems of NDAs, which will spell out stiff penalties for breaking confidentiality. In personal relationships, people who get married will be 'encouraged' to cooperate by the prospect of a difficult divorce. Even if we consider the rule of law, society attempts prevent undesirable behaviour by choosing to apply prison sentences or fines to people acting undesirably.

Cases where agents are in competition are more widely talked about in literature and, at least in my opinion, are considerably more interesting. One of the main patterns we will notice is that by being in competition and trying to steal utility from each other, agents can actually find themselves in a worse position than if they had tried to cooperate. Fans of Karma will no doubt be delighted by this. We will discuss a few examples of how situations like this can arise.

The first of these that we'll explore is probably the most famous and is known as 'The Prisoner's Dilemma'. In this scenario, two recently incarcerated prisoners (who know each other) are presented with the same options to ponder:

1. If you agree to testify against the other prisoner, you will get a reduced sentence of 1 year in prison while the other prisoner will receive 10 years
2. If both of you agree to testify against the other, you will get 7 years each
3. If neither prisoner agrees to testify against the other, the case against you will fall apart and you will only receive 2 years each

Here, if both parties could guarantee cooperation, they would walk away with the minimum total number of years in prison, at just four years. If one betrays the other and the other doesn't, the total number of years in prison is eleven and if they both betray each other, they will spend a combined total of fourteen years in prison. Clearly the best thing for both of them would be to keep their mouths shut and hope the other does the same.

Unfortunately, there's a problem. Let's look at the decision that each prisoner is facing. If they agree to betray the other, they will only have to spend one year in prison if the other doesn't betray him and seven years in prison if they do. If they choose to keep quiet the sentences are two years and ten years respectively, i.e., whatever the other prisoner chooses, they will spend less time in prison if they betray them. This is called the 'betray' decision 'dominating' the 'keeping quiet' decision, because whatever the other prisoner does, you're better off betraying them. To summarise the madness of this situation, we're in a position where both prisoners, working to further their own interests (competing), are likely to commit to spending the longest possible combined sentence in prison.

This kind of situation is something you will see happening all the time if you're looking for it. Let's say you and a friend are looking to buy a new car at the same time. You both have similar salaries, but you are worried that if you choose something cheap and they choose something more expensive, people will think they are more successful than you, which is a negative. Your friend is probably thinking the same thing. The cheaper car meets both of your needs and no one will think any less of either of you if you were both to own that car, but the fear of losing out pushes both of you to buy the more expensive car, which costs more and doesn't even give you the benefit of looking more successful. Ever noticed how well companies like BMW start doing in good economic times, even though (arguably) people's needs would be just as well met with cheaper alternatives?

Game theory is full of these kinds of examples, where seemingly bizarre behaviour can be rationalised by thinking about the decisions from the point of view of each participant. For another demonstration of this, you can consider the last time you found yourself driving down a road in desperate need of filling up with fuel. When you finally reach a filling station, you may have noticed, to your annoyance, that there are one or two others in very close proximity. How could this be good business? You will only fill up at one of those stations and had the others positioned themselves on the road you had just been driving down, you would have given them your money instead.

We can explain this phenomenon with a simple example. Let's start by considering two ice cream stands on opposite ends of a beach. Let's call them 'East' and 'West'. Both know that they are likely to get a stronger monopoly over ice cream if they are far apart, as customers won't be willing to walk the distance to the other, which means they should be able to charge more due to lack of competition. The first year's sales figures come in and both have done well; East has made a £10k profit that year, while West has made an impressive £15k. While both represent reasonable results, East, who has made less money, starts to envy their competitor's position. It looks to them as if there are a few more customers at that end of the beach. East decides that they are going to move their stall a bit closer to the middle of the beach. It should still be far enough away from West to keep a reasonably strong monopoly over ice creams, particularly at their original end of the beach. They might need to reduce his prices just a little to try and draw customers from the centre of the beach, who previously would have gone to West's stand, to persuade them to go to his stand instead.

Sure enough, the following year's sales figures roll in and East's profit has gone up to £12k. They're happy because they have improved on last year's figures despite having to reduce margins a little but unfortunately West has had a rough year, with only a £5k profit. They have suffered due to the loss of customers that are now going to East's stall. Note that the total profit made by both vendors has reduced because both have had to reduce their margins to deal with the increased competition (last year it was £25k but this year it's only £17k). West decides that East's stall is now in the slightly better place (more central relative to the rest of the beach) and if they want to go back to the number of customers they had last year, they are going to have to move a little nearer to them and drop their prices to compete. Again, this reduces the strength of their monopoly and further eats into the margins of each stall. This loop is likely to continue until both are sat right next to each other, probably making less money than they had originally but now neither feels like they are missing out on the opportunity for more customers. This is known as 'Hotelling's game' in the field of Game Theory.

The winner in all of this is the consumer, who can now afford more ice creams, as the competitors are having to slash prices due to the severe competition. The cost is that now they now have to walk a bit further to get it.

Of course, there are other reasons that businesses may choose to be close to others. One is association. Put your new restaurant between two others that hold Michelin stars and you might find you can charge a bit more for the food you would normally have to sell at market rates. You might also be able to get fine ingredients delivered for cheaper, as the delivery driver is already visiting that neighbourhood, or you could poach experienced staff from other businesses nearby.

This last point leads us on nicely to our final example. This is known as the 'The Ruin of the Commons', so named because it links back to when shepherds could take sheep onto common land to graze. For a while this works fine; the commons are large and there is plenty to go around so long as people don't start abusing the system. However, as soon as one shepherd starts to take more than their fair share by bringing more sheep, others are encouraged to do the same, for risk of losing out. Now the commons are overrun with shepherds trying to get the most they can out of the system and as a result the commons become depleted of all food, with the ecosystem having no time to recover. With everyone trying to exploit the system to its maximum, everyone taking part now has much less available to them than if they had cooperated and showed restraint.

Again, this has many real-life examples, the most pertinent of which is climate change. If all nations are competing, each will get a competitive advantage by breaking ranks and reverting to using a higher proportion of fossil fuels, as these will be cheaper than renewable alternatives (take the USA's position at the end of 2016). This will in turn encourage others to do the same. Should the cooperation hold, and humanity is able to turn the tide against this impending catastrophe, it will have been extremely hard won.

Think about this next time you stay late at work when trying to impress the boss. You may be inadvertently starting a cycle where everyone feels under pressure to stay later and later to earn kudos. Anyone not doing so will fear they will be overlooked next time the bonuses or promotions are getting handed out. This is likely to be pretty destructive, with individuals losing out on much needed time to relax in favour of staying late with everyone else, meaning no one stands out from the rest anyway.

All the above examples play out in the instances where individuals are acting entirely selfishly and considering only simple concepts, like time spent in jail and financial gains. All the agents involved will have utility preferences for their own gains, but they will probably also want to be liked and respected by others. There is also the question of trust that we discussed earlier. Let's take the Prisoner's

Dilemma that we analysed in our first example. While the game will play out as described if there are no consequences for being selfish, in reality you're probably going to make yourself quite unpopular with the other prisoner by betraying them, such that when they get out (whenever they do) you should at the very least expect them to have a bad attitude towards working with you again. If they have a particularly bad temperament (you have both spent time in prison after all), you can maybe expect a lot worse. The key here is that you don't know the Utility Functions of the other people you are competing with, so you don't know what success looks like from their perspective. All this adds considerable uncertainty to the games as they play out.

The key lesson is that if you want cooperation with other agents (as you should as much as possible), you will need to make the penalties for not cooperating severe enough to make the other agent think long and hard about breaking ranks when the opportunity arises. Obvious competitors are easier to deal with - If you are planning to set up a shop in an area that looks like you will have a strong monopoly, don't expect it to take long for a competitor to show up when you start doing well. In the words of a famous pirate from cinema "*Me? I'm dishonest, and a dishonest man you can always trust to be dishonest. Honestly. It's the honest ones you want to watch out for, because you can never predict when they're going to do something incredibly stupid.*"

Let's Change the Rules

So far, the impact of Game Theory on our environment seems quite negative. In the examples above, the presence of competition seems to erode our experience by leading to extra time spent in prison, lower profits and more expensive meals out. Are there ways we can spin this into a force for good? Yes, we just need to be careful about how we apply constraints to our systems.

Sometimes, small adjustments to the rules of the game can take an environment that appears intent on destroying itself and turn it into one where positive behaviour drives yet more positive behaviour. Take the example of plastic packaging and the impact this has on our planet, which was brought to our attention dramatically in the Blue Planet TV series, narrated by David Attenborough. After this documentary series was broadcast, and society became focused on the impact plastic waste was having on the Earth's eco systems, so much pressure was put on governments to legislate against waste of this kind that a Conservative government in the UK began implementing minimum charges on plastic shopping bags.

This began to steer the direction of competition in the market. Companies are now competing on who can provide the most sustainable products, with lowest impact on the environment. Supermarkets have introduced bags that not only adhere to the minimum pricing legislation but also are made from recycled materials and are longer lasting. This is Ruin of the Commons in its most positive form. Companies that cannot supply sustainable products are likely to find themselves out of business due to a negative attitude from society. Profits in these companies will probably be slightly lower than they would have been, but this surely cannot outweigh the positive benefits to the environment of this type of behaviour.

This type of system manipulation works in other areas as well. Currently in the UK it is required for restaurants and any outlets selling hot food to display their 'Food Hygiene Rating' (scored from 1-5) in their shop window and on any promotional material that they send out (mainly takeaway menus). Clearly the intention of the government is to improve the standards of every kitchen in the country by forcing them to display their scores for all to see. Any kitchen that thinks they will be able to maintain customers with a score of 1-2 will find themselves losing out to those with a score of 5 and similar prices.

In these examples, businesses can still choose how they operate. If they want to carry on producing plastic waste, they are free to do so. If they don't mind having poor food hygiene in their kitchens, then good luck to them! All that has been done here is the implementation of small pieces of legislation that can promote positive behaviour in these markets by changing the ways in which companies compete. Think of this as a 'Nudge' that sets off a chain reaction in a market by creating virtuous circle of competition.

I would like to think that this way of promoting change can be used to encourage desirable behaviour in other areas. Imagine if we gave companies scores like the food hygiene ratings mentioned above but based on how much tax they had paid in a country they were operating in, relative to what we would expect from their sales in this country. This score would then need to be printed in any advertisement that they ran in the country in question. Would this help to reduce tax avoidance in multi-national organizations? This isn't something I can answer but I would like to think it would encourage the biggest tax avoiders to 'invest' into a country's tax system to avoid an impossible task when marketing their products.

I recently had experience of an interesting problem that highlighted the phenomena that systems like this can lead to, and also offered the opportunity to explore how manipulating the rules can lead to desirable behaviour. This problem involved a distributed computing network that worked by gathering the idle processors on each individual machine and dedicating them to a central pool that could be used to run large batches of simulations. Everyone running the software on their personal machines simply decided how much resource they could spare and add it to the common pool; giving a lot of resource meant that these batches of simulations could be performed more quickly to the benefit of everyone.

The way this played out was fascinating. At the start, everyone seemed reasonably happy to give some of their resource to central pool, but this came with a compromise. Now anything you were running on your own machine would take a bit longer, as some of the resource was dedicated to running simulations for somebody else. Slowly people began to become a bit more selfish and only gave computing resource to the pool when they had sent a batch of simulations to the network themselves. Eventually they weren't even doing this. Batches of simulations now took much longer to come back and while everyone complained, no one was prepared to give their own recourse to try and fix the problem.

There were a few different things that were proposed as solutions to this problem. Measuring how much resource people gave over the course of the year and firing the person who gave the least was judged as being a little extreme (however I've heard rumors of practices like this that exist in other companies). Alternatives like dedicating a certain amount of computing resource by default as soon as you switched your machine on were thought of more suitable. This trick of changing the default behaviour is something that has worked incredibly well in other areas, like when asking people to join the organ donor register. In the end it was felt to be easier to buy some dedicated machines instead of trying to overcome the power of this effect!

Something we need to look out for when setting up systems like this is 'gaming' of the rules; that is, satisfying the letter of the regulations, but not the intent, by exploiting 'loop holes'. Going back to our plastic shopping bag example, let's say there is a regulation that states the carrier bags of a certain, named materials must be taxed by a fixed amount. Rather than make reusable bags, we could instead pour our efforts into research for materials that may have a similar, or worse, impact on the environment but avoid this legislation. We can now begin to use these bags and avoid the extra tax. What we've done is avoided the law, but not its intention, which was to reduce plastic waste and help the environment recover.

This kind of gaming of legislation is something I'm very familiar with in Formula 1, indeed it's something I believe the teams are world leaders in! When a new regulation arrives, attention in the team is immediately turned to how it can be avoided whilst still retaining the benefits of the now 'banned' system.

While this attitude remains, I believe the governing body has had some success in tackling this kind of gaming in two ways. Firstly, they have promoted an attitude that if a system is designed simply to avoid the regulations rather than their intentions, they will move to immediately have it banned. This discourages significant investment into areas where a team suspects they could achieve an advantage through rule avoidance rather than rule adherence. Secondly, teams are now forced to scrutineer their own cars, that is, self-certification for their cars against the rule book. Now, if a team is found to have a system that does not meet the regulations in the eyes of the governing body, who have been known to change their interpretation of the rules to ban systems they don't like the look of, they have a signed document from the team to say they have adhered to everything. I don't think there have been any penalties levelled against teams that have behaved in this way (which I'm sure was the intention) and I certainly wouldn't want to the be in the first team to receive one.

We will leave our discussion of Complex Systems here for the time being. I hope you have begun to see how having many, albeit simple, interacting components in a system can lead to some very strange effects indeed. We have left perhaps the strangest of all for our next chapter...

16

Butterflies and Hurricanes

Have you ever noticed that during news reports after a lunar or solar eclipse, the anchor is usually able to tell you with incredible precision when the next one is going to be? The same is true when a comet travels past the Earth. Sometimes these predictions are down to minute accuracy and the timescales can be over tens or hundreds of years. Now ask the same scientists what the weather is going to be like in exactly one month's time, and you will probably be met with a shrug.

There are certain things about the weather we can predict reasonably well from past experience; in July it is likely to be warner than in January (in the Northern Hemisphere). On a different scale, if it's sunny outside without any clouds, we know that we can probably afford to take a 10-minute walk without any risk of rain. However, imagine you were going on a weekend away and for some reason you had to decide which clothes to take two weeks in advance; I'm sure most of us would struggle with this and possibly end up packing half our wardrobe.

Is it because those involved in weather prediction don't understand the mechanisms that give rise to the different types of weather as well as the scientists studying the orbits of the bodies in space? No. This difference in predictability comes about because the weather system includes effects that make it *inherently* unpredictable. The systems of our climate are built such that they defy any attempts to understand how they will appear in the future, even when using the world's most powerful computers and sophisticated models. Think of this as nature's way of keeping things interesting. It is known in mathematics as 'Chaos', and we'll see that the weather is far from the only system that behaves in this way.

Of all the phenomena we will describe in this book, Chaos must be both the strangest and the one which will affect our task more than any other. As a subject, I find it completely fascinating that we can create an incredibly accurate model for a system that is basically useless for predicting how the system will look beyond a finite time 'horizon'. As a species, we have sent people to the moon, cured some of the deadliest diseases and created huge networks of infrastructure, yet we find it basically impossible to know whether we will need to take an umbrella with us to work in a week's time. This theory states that even if we could treat the universe as deterministic, there will still be things in our future that will surprise us. After all, what's the point of reading the story if we know how it's going to end?

We will use the term 'chaotic' very deliberately to describe systems with relatively simple rules that can diverge into any one of an infinite number of different directions. Something we should be clear on is that Complexity is neither a necessary nor sufficient condition for chaos. There are Complex Systems that do not behave chaotically and there are chaotic systems that are not complex, but we should definitely be aware of both when we are considering the environment. Let's take what I think is one of the simplest chaotic systems; a double pendulum. A single pendulum is a very predictable system; indeed, this has been used to keep time when building clocks for centuries. You might imagine that if you were to add a second pendulum onto the end of the first, it would behave in a similarly predictable way. However, this system now starts to behave very strangely. There are many videos of systems like this online and I highly recommend that you search some of them out. I am very fortunate to own a desktop ornament that is based on the same principle, and I find the movement completely captivating. Its presence has probably delayed many pieces of work that I could have finished much earlier if it wasn't for this distraction. The 'periodicity' of the pendulum system, or the tendency of the pendulum to behave regularly over time, is completely broken down. The pendulum flails around in a seemingly random way, with the positions of each apparently completely independent of their positions a few seconds ago. The system never seems to repeat itself, with each movement looking like it has happened for the first time.

Figure 13. Double pendulum desktop ornament

Something you can try to do if you get your hands on one of these is to try starting it off twice from what you imagine is the same position. You will find that even if the motion starts off looking the same, it quickly deteriorates into something completely different. Here we hit upon one of the defining characteristics of chaotic systems; the sensitivity to initial conditions.

This sensitivity was initially discovered during very early computer simulations of weather systems. On one occasion, Edward Lorenz, an early pioneer of this theory, came back to the computer he was using only to discover it had stopped working during the night. He took this for an annoyance rather than an indication of anything serious; he would simply restart the simulation from a few seconds before it failed and carry on from that point. Unfortunately, after restarting the simulation it started behaving in a completely different way to the original. After some confusion, this was eventually traced to the fact that he had only entered the initial conditions accurately to a few decimal places. These tiny errors, that in a non-chaotic system would have made next to no difference, were enough to cause the system to travel in a completely new direction.

Figure 14. The 'Strange Attractor' describes a simple chaotic system in 'phase space'

This is linked to the famous 'Butterfly Effect', which describes a possible scenario where a butterfly flapping its wings in Africa can change the behaviour of the air enough to cause hurricanes in North America. The initial conditions are said to 'explode' with actions at tiny scales influencing them at the very largest. We should be clear that this is different from instability. Imagine a ball sat on top of a tall hill, nudging it in either direction will cause it to gather speed indefinitely, leading to it reaching incredible speeds. We cannot describe our climate as unstable, as it has broadly been the same for centuries. Extreme weather events come and go but we cannot necessarily say we are on an accelerating path to the extinction of our species.

Despite this, the Butterfly Effect is going to cause big problems when we're trying to predict the future. If we are to model what will happen at all scales in the system, we will need to be able to measure the behaviour at the very smallest. Even with our current models for weather, increasing the accuracy of our measurements for the initial conditions (say temperature, wind speeds, atmospheric pressure etc.) by factors of hundreds, thousands or millions will only be able to increase our prediction horizon by a day or two. There may be certain conditions that we will find slightly easier or slightly harder to predict, but this characteristic is always likely to exist.

We should expect similar results in other chaotic systems, such as movements in the economy, which is something that people who are trying to make money from you will claim to have the power to predict. Again, we must be mindful that our ability to see into the future is a function of the nature of the system, which in most cases is likely to be very chaotic indeed.

A cliché that appears in many 'feel good' movies is that one person has the power to change the world. Indeed, there are several examples of this in history. Individuals (albeit probably with assistance) have discovered vaccines that have eradicated diseases across all of humanity, have made inferences about our reality that have led to changes in the way we approach science, and invented products that now exist in every home in developed countries. For me, this is an example of the Butterfly Effect, or actions at the smallest scale affecting the largest; you could imagine that the present would look very different if there were small changes to the initial conditions in these examples. Maybe the people described above chose a different career path? Maybe their parents never met? Maybe they were hit by a bus the week before stumbling upon their brilliant idea? A more sobering thought is how many advances have we missed out on because the conditions for their creation were only slightly wrong.

While one person may be able to change the world, chaos suggests that one person is very unlikely to be able to predict the future, at least not with any regularity. We will frequently hear forecasts that describe how an individual or group thinks something will look in a number of years' time, be it a company, a career, a relationship or anything else. These forecasts may be meant with the best intentions and will serve as a useful guide, setting targets for what those involved should be aiming for. Indeed, having a difficult target to achieve might lead to some additional creative thinking that can benefit those involved immensely, but unless they capture everything the environment can throw at them, these forecasts are likely to be pure fantasy.

As an example, Formula 1 is a competitive industry where people are not happy to be losing for years on end, therefore the push to improve is captured in the forecasts for how the team bosses think we will perform in the future. I have frequently been in team briefings where people with the best of intentions have described how if we keep on pushing as we have been doing, the team is going to start winning championships in 'x' number of years' time, be it three, five or any other number. The thought I always had when listening to this is that there will be nine other team briefings around the world occurring at similar times saying *exactly* the same thing. Given that only two championships are handed out every year, there are going to be a lot of disappointed people.

In these teams, we are not only competing with the environment that includes rule changes, financial disturbances, politics and much more, but we are also up against nine competitors that have very similar working practices, similar expertise, and the same desire to win. How on earth is anyone able to predict with any confidence that they will be able to beat them all in some number of years' time? These forecasts can be made to look more believable by having a slow climb to the front of the grid, maybe with a year or two as third or second fastest team before the glorious run to the championship. Unfortunately, history is full of examples of unforeseen events that conspire to ruin your extremely neat forecast for how you expect things to play out. These events can happen at the most dramatic of scales.

In 2008 there was a huge stock market crash, when banks who had lent money to people they shouldn't have started to realise that they weren't going to get much of it back. This had effects across the world in industries that had little to do with banking (as most industries have quite a lot to do with money). Many people lost jobs, businesses closed, profits became losses. This was something that took years to recover from.

The Covid-19 outbreak is another good example of something the environment can throw at you that completely destroys any forecasts you had made that covered the time during and after the outbreak. I would like to see a forecast from the year before that said you wouldn't be able to stand within two metres of somebody when shopping through this period.

It's not all negative though. In 2016, in one of the biggest shocks in the sport's history, Leicester City FC, who had been tipped by many to be relegated, won the Premier League title, beating off some of the biggest clubs in the world in the process. Their success was all the more remarkable due to the length of time they were able to sustain their performances. This is not a simple knockout competition where you only need to perform well in a few key games. This was a league where each team plays thirty eight times, playing every other team both home and away.

These examples demonstrate that the world is a very chaotic place. It is full of surprises that we definitely need to bear in mind when predicting how things will play out. Given the complexity of the problems we are likely to be dealing with, I would advise that we steer clear of dramatic, crowd pleasing forecasts an instead lay out multiple plans for how we are going to behave should certain scenarios occur. This way, we can refine our course depending on the outcome of events that we have no way of predicting. If the only forecast we can make is to try and do the best job we can in all circumstances, that might be the one that's closest to the truth.

17

There's Something Holding Me Back

How is it that I find myself stood outside a railway station in the middle of the night, in a town I didn't recognise, with no idea how I'm going to make the thirty-mile trip home? I'd been to see a band in London with a friend, and while the plan for the return journey had sounded so simple in advance, the reality had been somewhat different.

Yes, the gig finished in plenty of time to get the last train home, but it turns out that when you have tens of thousands of people all with the same idea, the process takes a little longer. I remember us both stood in line for the nearest London Underground station, staring at our watches, convincing ourselves that we could still make it to station in time. There were police trying to keep the liveliest members of the queue from carrying on the gig outside, and the atmosphere was a little tense.

Eventually, we got to the front and onto a tube train going in the right direction. We were talking to another group who'd been in the same queue and they had sounded convinced that they would make it on the last train to Reading. This came as a relief, as this was the line that our station was on as well. Unfortunately, when we arrived, it turned out that Reading was as far as the train would go, and getting to Didcot wasn't going to be possible. This was great news for our new friends but came as extremely disappointing news to us.

We got on the train regardless, as we thought we'd rather be closer to home than not. When we arrived at the last stop it was around one in the morning and we knew our only real option was to convince a local taxi driver to take us on the sixty-mile round trip to the station we were originally aiming for and cough up the cash. We could have probably avoided the most expensive taxi ride in history if we had appreciated the following constraints of London's transport system:

- There isn't enough car parking in the centre of London for the number of people that would be at the gig.
- You can only fit so many people on a tube train and the extra passengers will have to wait for the next one.
- It's difficult to get a train anywhere you want to go at midnight.

Whatever systems we encounter as part of the environment, they are likely to involve constraints that will limit our ability to maximise the outcome of our task. We have already come across other examples when describing the inputs and outputs to our driving system. Our ability to accelerate and decelerate is limited by the capability of the car and we should probably avoid exceeding the speed limit significantly, unless we're happy walking to work in future.

When we talk about constraints, we are talking about limits on what will be possible in the system we are in. We will put these into two categories:

- Natural constraints
- System constraints

The constraints that arise from the laws of nature are not things we are going to be able to do anything about; to name some of the most popular, we can't have perpetual motion, we can't travel faster than light and we can't travel back in time. These of course sound absurd in most circumstances but maybe only because our experience with real life (maybe in a parallel universe the speed of light is walking pace?). In the world we have built around us, we're pretty unlikely to encounter many of these in our task planning, unless we're designing some incredibly interesting machines.

Other kinds of constraints that arise from the natural world cover things like the strength of our materials and the extent to which our resources can be used to help accomplish our goals. This is something that those developing F1 cars will encounter when adapting their designs to survive accidents. If we want to avoid fatalities, nasty injuries and excess damage to the internal components of the car, we must keep accelerations below a certain threshold during the accident, as well as ensuring we don't get any nasty deforming of the cockpit. These acceleration limits will be set by the ability of human body to survive intact during impacts. The length of crash structure we need to keep the acceleration at the desired maximum will be dictated by the speed the car is travelling. Above a certain speed, the length of the crumple zones required become impractical, meaning we will be unable to meet the deceleration target at that speed. The strength of the materials we are using will also play a role; if we have a heavy car to slow down in a short distance, we are going to need a lot of strong material to do it, which of course will add more weight to the car, making the problem harder still. All these considerations would be change if we had access to a different set of materials or we were dealing with different laws of nature.

System constraints, on the other hand, will be constraints that exist because of systems we have limited or no control over, and are something we are likely to come across much more frequently. The train timetable is fixed by the rail operator and there will be nothing we can do about it, no matter how much we want to catch a train at 3 am. There is no law of nature that states we cannot have a train that goes where we want at that time, but it doesn't mean you'll be able to get on one.

These constraints can be static in their appearance or change over time. You could imagine going to a specialist dealer to get a replacement part for your car. When you arrive, you are told that unfortunately the customer before you had just bought the last item they had in stock. Here is a situation where we've gone from having no constraint on being able to purchase this component to it suddenly not being available from that outlet. The nature of constraints can change all the time, meaning we need to be aware of the possibilities when modelling our task.

We can think of constraints as either being 'hard', that is, impossible to overcome, or 'soft', possible but inefficient to overcome. All our natural laws will fall into the hard constraint category, while our system constraints are likely to be described as soft. Think about the problem of getting from the UK to the USA on short notice. It's likely that all the major airlines will be sold out, making buying a ticket through conventional means impossible; but that doesn't mean we should give up immediately. If we know someone who has a ticket, we might be able to purchase it from them at an inflated price, or alternatively, there will be private operators that should be able to charter a plane for a sizeable fee. Given enough money there will surely be a way to accomplish what you're looking to achieve. It will then be up to you to decide how much what you're looking to do is worth to you personally. In most cases however, it will be appropriate to treat soft constraints as hard, particularly when the cost of breaking them is spectacular.

Some constraints are likely to be more 'active' than others, that is, having a larger impact on the process. The time taken to deliver a package to another location is likely to be measured in days, if not weeks. These kinds of delays are something that we must be very aware of when framing a plan for how we will proceed with a task that will run over similar time scales. On the other hand, the time to deliver an e-mail is measured in seconds if not less, which therefore is less likely to influence a task that spans weeks or months; although it could have an impact if we're trying to make decisions in a very small space of time.

18

Order From Chaos

All of this might feel like a lot to bear in mind when we're making models of the environment we are present in. Complexity teaches us that the environment is likely to be made up of many interacting elements, which may have simple behaviours on their own, but when grouped together give rise to new phenomena that cannot be explained by looking at each individual component. If the interacting parts are able to adapt to the environment as they see it, we can get very strange effects that can be traced back to principles of Game Theory. In a chaotic system, it is possible that small changes in our initial conditions, which will be impossible to measure, will have a dramatic effect on our outcome even over short time scales. These effects are going to cause us big problems when we are trying to manipulate the future in our task. There are however some things that we can bear in mind when it comes to deciding how to behave in the face of these difficulties.

When modelling, we must be mindful of the 'scales' we are interested in; when we're estimating how long our sailing trip is going to take, we are not interested in the local micro-currents of the air; we are more interested in the overall wind speed and direction. This emergent property is likely to be much easier to predict than how every air molecule is interacting with the environment and others around it. The simplest model that captures what we're interested in should always be the most accurate and the most useful model we can have. This will be relatively efficient to produce and run, and won't give us information we don't care about.

There is a fallacy in science and engineering that if a more complex model of a system better captures the internal workings, it will be more accurate. This is not true, as for every new modelling parameter we add, we are introducing extra uncertainty. If we can group the levels of the hierarchy we're not interested in together, we are far more likely to be able to recreate emergent properties of the system and give ourselves more powerful handles with which to drag results around, which will be useful if they are not behaving as we expect. In terms of the system hierarchy, think of adding extra complexity as moving down through the levels, further away from the effects you're most interested in.

Let's go back to our F1 car model example. The physics at work in an F1 car goes down to the atomic level, with chemical bonds being broken and formed during combustion in the engine and as part of the generating grip in the tyres. Does this mean that we need to understand how these processes work at this scale if we want to know how to get the best out of it? Not really. Man had fire long before he understood combustion at a chemical level. Our ancestors' 'models' suggested that if they took something that was already on fire and held it close to something that wasn't, that object might catch fire as well. This is the modelling of the emergent properties of fire, rather than the chemical reaction itself. We can take a similar approach to our car model; if we know how a new engine behaves as a whole, without necessarily knowing how the combustion unfolds in the combustion chamber, we can get a very good approximation of the effect it has on the track. The same can also be said for a suspension device or part of the aerodynamics.

We need only go into more depth and detail when we're interested in the results at those scales. For example, a professor of chemistry who's interested in how polymer chains interact will need a rubber model at a finer scale than someone who is only interested in how much grip a tyre will have.

Of course, it may be too simple to dismiss the workings at lower scales when modelling a system, as it might be the interactions between different levels we are most interested in. We may also benefit from having a single model that works across different scales, as we will no longer have to maintain multiple models and correlate them with each other. Now we will run into the problems that we have already discussed. We can pick a scale for which we will correlate reasonably well but now scales below and above will start to drift. This is expected, given the complexity of the interactions and how these lead to the emergent properties of the system.

Something we can do is add 'fudges' throughout the model to make sure the correlation at each individual level is as good as it can be. Let's look at our car model example again (as this this is a problem I come across regularly). The tyre team will usually be in charge of the tyre model that is used in the full vehicle model. This will be based on their own measurements and they will be convinced that it's as accurate as it's ever going to be. Similarly, the aerodynamics department will have their own model that they believe explains the forces that arise from the aerodynamics as well as they can. We will then have models for the suspension, steering, powertrain and chassis. When we put all these together, we find that this model doesn't really behave like the real car does. The emergent properties, like the 'handling' and 'performance' of the car do not match what we measure from reality, which is a problem because, ideally, we'd like to use the model to predict how they will affect these areas from the starting point we have understood from the track.

This problem is puzzling and is a consequence of a reductionist way of thinking about modelling. Why should it be that just because we have all the component parts working in isolation that we should see the whole system behaving as we expect? The interactions *between* the elements should concern us as much as the elements themselves. We could choose to target only the emergent properties, like the handling and performance of the car, and we are likely to get much closer, however this ceases to be useful because we can't see how changes lower down the hierarchy of complexity (the suspension, aerodynamics or tyres, for example) might affect overall performance.

The solution is to have a 'fudge' layer between the components of the model and the emergent properties. These are forces that are applied to the car which *must* be there to create the handling/performance we observe on track, but we cannot account for them through the sum of the constituent parts, which avoids the need to adjust any measured data we have at system level to achieve the desired effects higher up the system hierarchy. We now also have a measure for how successful we've been in modelling the sum of all the elements of the car. The larger the magnitude of the 'fudge' forces, the worse the total correlation of the constituent parts must be; when we manage to reduce the magnitude of these forces, we know we're on the right track.

We can consider similar fudge layers between systems and subsystems. Let's say we have a very sophisticated model for the tyre rubber interaction with the road but when we go to the level above, it doesn't really have the grip characteristics that we think the tyre should. We now need a to create some forces that capture how we think the tyre behaves and create another set of corrections so that now the model does what we want as a whole. The only thing to bear in mind is keeping track of these corrections, so no other measures are brought in over the top of them to undo their effects.

Eventually we will run out of knowledge when it comes to a particular system and reach a point where we cannot parameterise a certain area further than we have, which is similar to the 'incompleteness' we've discussed in previous chapters. This could be because the element in question is a confidential part of a competitor's system (like their Utility Function) or we don't have the means to measure it. At this point, we can run simulations of what different values of this parameter will do to our system; we might even be able to infer what this parameter may be based on previous results for this system. We can then decide whether we care about it or not. For example, we may not know how much lunch is going to cost at the restaurant in the hotel we're staying in but if we know that there is a supermarket a few minutes down the road, we are unlikely to be short of somewhere to eat during our stay. If this were not the case, we can make sensible predictions about how much the food is likely to be and come up with plans accordingly.

In a similar vein, we need to be aware of the parameters we are extremely sensitive to in our system. In our driving to work example, this could be the magnitude of the traffic at that time of day and whether there are any roadworks on route. This allows us to formulate plans for multiple scenarios, should those things occur on our journey.

Again, the point is not to have all the answers about how the future will pan out, as this requires knowledge beyond what we can obtain. Much better is to have all the plans that allow you to adapt to things that cannot be predicted.

Final Thought on Modelling

Our models will capture our complete understandings of the problems we are dealing with; they are the sum of all our knowledge of a particular system and we should keep them in good shape. If we believe that a model doesn't behave in the way we think the system it represents does, this is not an argument for dismissing the model; it is instead an argument for modifying the model until it behaves how we would like. I have been surprised on many occasions by some people's desire to dismiss the models they have been using because they do not follow their own pre-conceived ideas of how the system should behave. If you find yourself in this position, there will always be some kind of experiment you can do to demonstrate that the model is incorrect, at which point you can use the results from this to update your model. Alternatively, your hypothesis is wrong.

In my experience, finding a justifiable way to change a model so that it behaves more like reality is difficult. Chances are, all your original understanding, like the laws of nature and physical measurements you made, were already included in the first version. Now you're faced with breaking this order with a 'hack' or fudge that will look very ugly. For me, making this change early means you can benefit from the increased accuracy of the model sooner, while you are debating what could be wrong with your initial assumptions to cause the poor correlation. Remember, you won't be rewarded for having a nice clean model that can't be used to predict how the system it represents will behave in the future.

The existence of chaos leads to the question of how we should treat something that we are going to find difficult to predict, which links back to the system output uncertainty that we discussed previously. If we're looking beyond our prediction horizon, we do not know what the environment is going to look like with certainty, but we probably know what the best and worst cases are going to look like. The growth of the economy is unlikely to be lower than it has ever been and unlikely to be higher than it has ever been; looking back through historical trends is likely to alert us to any cases where one is more unlikely than the other. For example, given the trends in global temperatures, July 14th 2035 is more likely to be warmer than July 14th 2015 but the difference is unlikely to be 100 °C.

This gives us something to anchor our planning to. Hopefully you will not find this to be groundbreaking in its power. When faced with the uncertainty over what the weather will be like in the second week of our holiday, that doesn't mean we give up and cancel the whole thing; we simply need to pack for a variety of scenarios, almost guaranteeing we will have something suitable for whatever the weather. Now we need to start applying the same logic to other situations we know to be uncertain.

19

The State of the Environment

In the previous section we described the agent as having a 'state' at any point in time within the duration of the task, which contains all the information to define the properties of the agent. We can say the same for the environment, at least for the elements we are interested in. This state will be all the information we need to uniquely define the conditions of the environment at that point in time.

Unfortunately for us the precise state is unknowable, and we can only infer what it might be using our imperfect measurements. We can, however, still define what it should consist of. A prediction for the state of our environment is likely to be a lot more useful than having nothing at all.

The difference between the state and the output of the environment is only in the number of parameters we are measuring; the output contains the things that we can measure from the environment, which are likely to be of most importance to the control process. The state is likely to contain extra information that is necessary for the internal workings of the system but is either not possible to measure or not interesting from our point of view of modelling the task.

Let's take an example of a competitor car in a race situation. Our car is ahead but the following car is reeling us in fast. The output that we can measure is the gap between the cars at each timing point around the track. This is clearly important to us - too close and they will try to overtake. What we will not be able to see is the internal 'state' of the other car (for this we would need a model). This would contain information like whether the driver has pushed too hard on their tyres, which will now start to degrade and require changing, or whether the car is using an engine mode that gives more power and they will only be able to run for one or two laps. Maybe the driver has made a different change to the car that has unlocked some performance they hadn't been able to use up until this point.

Our ability to determine what we do next will depend on the accuracy of our model in this situation. How confident are we that the pace of the competitor will only be short lived and they will eventually give up the chase in order to get their car to the end of the race? Or maybe this is just an occasion where we need to prepare to be overtaken? In either case, our predictive ability is likely to be riddled with uncertainty as a result of some of the phenomena we've described.

20

Environment Summary

Over the course of this section, we have discussed many different aspects of our environment and how the systems within it can behave. We've described how the environment is the stage on which our task will be played out, and the fact that it is made up of anything we are not in direct control of as the agent of the task.

We can expect our environment to be made up of many systems, the nature of which will vary depending on what it is we are trying to do, but they will either be natural or man-made. Some of these, like the systems we work with in our jobs, we will be able to influence, while others, like the weather, we will not. We'll constantly receive outputs from our environment, which can come in response to any inputs we have made, or they could come as disturbances that cause our task to change course without warning. We won't be able to perfectly interpret the inputs we make or the outputs we receive from the environment, which will make the exact state of the environment unknowable to us - although we should be able to estimate it.

The environment we inhabit is likely to be a Complex System, made up of a hierarchy of interacting elements that give rise to emergent properties which cannot be explained by only looking at the bottom layers in isolation. Complex Systems can either be physical or adaptive; complex physical systems only rely on natural laws to explain their inner workings, while adaptive systems involve feedback loops that can change the behaviour of the system as it evolves.

Our complex adaptive systems can include many different agents, who can either be cooperative or competitive. We should be able to predict the behaviour of these agents, and therefore the system as a whole, by understanding the teachings of Game Theory. Cooperation between agents is usually unstable, as trust can be broken much more quickly than it can be gained. Games with competing agents can often lead to scenarios where, by trying to take utility from each other, the whole system deteriorates to a point where the competitors are receiving far less than if they had cooperated. In systems we have influence over, we should be able to manipulate the rules of these games to promote derisible behaviour.

Many systems exhibit chaotic behaviour; characterised by a high sensitivity to initial conditions, which can lead to unpredictability. In these systems, disturbances at the smallest scales can give rise to changes in the emergent properties at the highest levels and we should respect the prediction horizons when trying to make forecasts when chaotic systems are involved.

Our environment will include constraints on the way we can behave, which can be separated into natural constraints and those that are the products of our own systems. We can think of these as hard to avoid, or soft; simply inefficient to avoid.

All this information can help us to understand and model our environment. Our most accurate models will be the ones that only concern themselves with the levels of the complex hierarchy that we are most interested in. If we want to understand how the different levels interact with one another, we can use 'fudges' between levels of the hierarchy to ensure we get correlation with reality across all levels. We must be aware of the parameters our correlation is most sensitive to and ensure most effort goes into reducing the uncertainty in those

Once we have the models of our system, we should follow their guidance unless we are able to demonstrate that they are incorrect through experiment. In these instances, we must improve the model to capture the new behaviour, rather than dismissing it as untrustworthy.

We have now described how both our agent and our environment are likely to behave during our task, but we are missing the element of the model that can connect the two. This will be a method for manipulating the environment to achieve the desires of the agent. This process we will refer to as 'Control'.

Part Three

Control Process

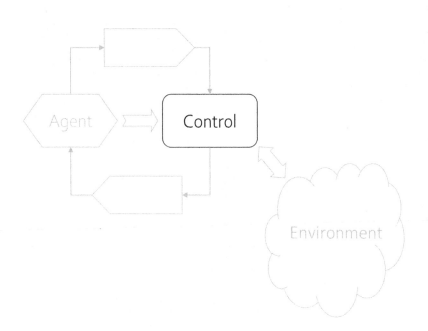

21

Living the Dream

My first few months in Formula 1 weren't quite what I imagined. My job had originally been advertised as a software engineer but when we were told that the team was building a new simulator, I volunteered my expertise. Flash forward a few months and I found myself in the abandoned operations room, in the middle of winter, with no central heating, trying desperately to get computers to talk to electric motors.

The computer I was using for this task had been deemed 'too noisy' to stay in the office and a new 'lab' was created for my work away from where I could disturb other people. Whilst I was getting to grips with the software we had chosen to use, I had a single electric motor that I could use for prototyping. The final design was to have many, much larger, motors that would all need to work together to move the simulator cockpit in a way that would tell the driver how the car was handling. Watching this single motor slowly rotate around couldn't have felt further from that.

A few months after that, the first parts for the new platform began arriving and we started to assemble them in the room where it would operate. This phase now seemed much more impressive. All the hours in the cold, lonely ops room had at least been productive, with the control system working reasonably reliably from the off. One of the most interesting parts of developing the control system was running the new simulator platform behind the old one during a simulator session; you could then watch the platform move in response to the outputs from the model, without a driver in it. While this was a lot of fun, the fact that a human being would be sat in the machine I had been tasked with controlling had not escaped me. Not long after that first human being, it would be one of the race drivers, which the team would probably like to keep in one piece.

There were a couple of hiccups, but the first test day eventually came. There had obviously been a lot of checking and re-checking of computer code and safety systems but I was still somewhat anxious. The first few laps went happily without incident and the feedback we got surpassed all our expectations. We got glowing reviews from the drivers about how it transformed the feeling of the simulator, allowing them to drive it in a way that was closer to how they would drive at the track. All this with essentially the first version of the code.

We had taken the requirements of the driver and produced a system that could achieve them within the constraints of the environment we were operating in. This is the process of 'control' and it's what we're going to talk about in this section.

22

In Complete Control

Now we have introduced the agent and their environment for our task, we have the desires we need to satisfy, the resource we can use to do it and the situation we are going to try to achieve it in. What we are missing is a mechanism for the agent to act on the environment to achieve these desires. This is what we will describe as the 'Control Process'.

When you want to travel somewhere in your car, what do you do? You get in your car, turn on the ignition and position yourself in a way that you can operate all the controls. Then, through a combination of practice and intuition, you can usually navigate successfully to your destination. The steps required to do this are phenomenally complex (anyone who has experience of the development of driverless cars will understand just how complicated this problem is). You must interpret the conditions on the road whilst simultaneously controlling the trajectory of the car. You must react to unexpected disturbances, like the road you want to travel down being blocked, or someone crossing in front of you without checking for cars first. Learning to do this takes many hours of practice but, given enough time, it becomes second nature. This is an example of a 'Control' problem; you take what you want to achieve (to reach your destination) and turn it into a series of steps that act on the environment that are intended to achieve it, including any reactions to unexpected events. In other words, 'control' is what links the agent to the environment when trying to complete the task we are exploring.

We will use the term 'Control' to describe how the agent is able to affect the environment using their resource and understanding to maximise their utility. While we'll be using elements of control theory to describe how this can be achieved, this definition is clearly broader than just this subject. In many cases, our 'Control Law' will be very simple; if the room is too hot, we turn down the central heating to a level we expect to find more comfortable. If we want to make a quick meal, we can put a ready meal in the microwave and follow the instructions written on the side of the box. If we want to send an e-mail, we turn on our computer, select the right program and begin typing. However, we can also use this term to describe the steps we will usually go through to bring a product to market, build a skyscraper, or drive a car.

Granted, each of these processes contains a significant number of smaller steps that require more complex control actions. Considering our central heating example, the process of getting up out your chair, moving to thermostat and turning the dial to make the temperature more comfortable involves a massive number of movements and measurements in your body. This is not unlike our Complex System that we introduced in the previous section. While there may be an entire hierarchy of control actions involved in the processes we are performing, we only care about the actions of the agent at the scales that we normally operate on. This is usually the most efficient way of describing each step.

What we want to capture is how you would normally behave towards feedback that you get from the environment when performing a task. If your product is getting poor reviews on industry websites, how would you typically act - can this be boiled down to a few simple statements? What about if the new product is doing so well that it's not possible to produce enough to satisfy demand? Or if there is a fire in the warehouse that means your stock is depleted? In these situations, you *will* react somehow (because your utility will drop if you don't), and this section describes how we can capture those reactions into something more formal.

The steps involved in the control problem can be summarised using the diagram in Figure 15. Our targets arrive on the left-hand side as an input and these are compared to our current estimated 'state'. If our estimate doesn't match our target state, this difference is transformed into inputs to the environment using our control law. These inputs get fed into the environment, where they are turned into outputs via the systems inherent within it. We then need a process of measurement to interpret whether we have achieved what we expected. This produces information that is returned into our updated state estimate, and is subsequently compared to our targets at the next time step.

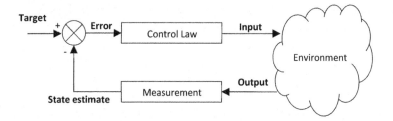

Figure 15. Illustration of control problem

If we take our driving example, the target is to reach our destination, our control law is to take our desired position and trajectory and turn it into a set of operations on the car's controls. The environment is the car and road network. Our measurements will be made using the cars instruments, our own perception of the conditions on the road, plus any navigation programs we might be using. If we find we are not making the progress we were hoping for, or indeed if we are likely to arrive to early, this is fed back into the control law and new inputs are created.

There will be costs involved in applying actions to the environment, for which we're going to need some of our agent's resource. In the case of the car journey, this is fuel, our time and probably some physical exertion. In our industry examples, this is going to be financial resource plus the time and skills of our employees. These actions will have a cost but provided our utility is increasing (or at least not falling faster than it needs to) we can see this as worthwhile investment.

If we're performing a task involving some sales or similar, we may also get some resource returned as part of the control process. If we are selling items in a shop, we will accumulate revenue from every sale in exchange for our resource of stock, which can go straight in the bank to be reinvested if necessary, or passed to shareholders in dividends to increase their utility. The allocation of this new resource will depend on the Utility Function of the agent we are dealing with. More on this in a later chapter.

Content of This Section

In this section, we will discuss each step in the control process. We will start, perhaps surprisingly, with the important process of measurement. We're already familiar with the targets set by our agents from a previous chapter, but we won't know how to behave without first trying to understand the current state of our environment. We may *want* to spend a nice day at the beach, but you wouldn't start packing the car with all your beach clothes without first looking outside to check the state of the weather.

Then we will go onto to discussing control laws and the forms they can take, thinking about the difference between discrete, flow chart-like approaches to our control law and to continuous, 'regulator'-type problems. We'll then go on to combinations of these approaches and how they each contribute to the success of the task.

Another action we can choose to take is to perform experiments on the environment. These will probably not have any direct influence on the outcome of our task (other than to take resource away from other areas) but it will help to reduce the uncertainty we might have that is causing us problems in selecting our control actions; there are important considerations we must understand when designing any experiments.

We've already discussed the constraints we may encounter in the environment as part of the previous section. These constraints will also feature in the control laws we design which we'll dedicate some time to thinking about how to approach these.

Finally, as with all sections in this book, we'll discuss the uncertainties involved in the control process and how these can affect our choice of actions.

23

No Signal

The targets we get from our agent are one of the elements we need when deciding how we are going to behave in any given situation, and the other is an estimate of the state of our environment. These estimates can only be made once we've made some 'measurements' on our environment's outputs. When we say measurements, we shouldn't only think of those that can be obtained from instruments that a scientist or engineer might use; we can also use surveys, our senses or even just responses to questions we ask our friends. All we're looking for is something that can convey useful information from the environment to us in some way.

The measurements we take on a Formula 1 car are essential for making sure everything is working properly. Indeed, without the measurements in the throttle and brake-by-wire systems, you wouldn't even be able to go anywhere. A great deal of investment made in ensuring we have the right sensors for every task, which isn't just the most accurate but also the lightest, least intrusive and most robust. The information recorded from these is studied by the team of engineers at the factory and is vital for understanding the limitations of the car and what is necessary to make it go faster.

In our task, there could be any number of different options for measurements we can take. We will steer clear of deciding what kinds might be better or worse in a particular situation; that will depend entirely on what we're trying to achieve. The steps we go through when deciding what measurements to take, however, is something that we can talk about here, as the quality of the measurements we get will dictate the quality of inferences we can make. This step may well involve processing of these measurements to give them more meaning before we can use them - we'll come onto how we use the results to make inferences on reality in the next section.

Most of us will be familiar with the process of measurement. Driving a car would be a lot more difficult without a speedometer to tell you whether you're going too fast. When you're putting up shelves, you can rely on a spirit level to tell you whether your books are likely to slide off, and somewhere in the back of a drawer you probably have a medical thermometer for taking your temperature when you're feeling unwell. We can also consider things like surveys, often delivered by those people you try and avoid on city high streets, as forms of measurement.

We happily use the values we recover to draw conclusions about the state of the environment, without fear of inaccuracy. What is often overlooked is that these measurements are not reality; they are only estimates that we have made using the tools we have at our disposal. Indeed, unless we're dealing with pure mathematics, 'reality' is going to be unknowable.

Our measurements have the potential to lead us astray in many ways, as we'll think about under the headings below.

Relevance

Before we consider the errors we might encounter when making measurements, we must first consider the relevance of the measurements we could take. This is an interesting problem to face, as sometimes the most relevant measurements we can take are difficult or expensive to perform, while less relevant, but simpler measurements might still be informative. Let's take the example of measurements used in health. When assessing whether we think somebody is overweight, there are many different techniques we could use.

The first measurement we could use comes straight from weighing scales. We know that someone weighing 130 kg is unlikely to be a healthy weight, however if they are very tall, the problem won't be as serious as if the individual is below average height. We could then move to a slightly more sophisticated measurement technique, such as BMI or Body Mass Index, which takes height into account as well. This measurement will therefore tell you whether you are a 'healthy' weight relative to the expectation for your height. But we still have problems. I'm sure you've heard examples of individuals who would normally be considered very healthy being labelled 'overweight' using this metric. The fact that muscle is relatively dense and therefore heavier than fat means the measurement is still not completely relevant when deciding whether someone is carrying an unhealthy amount of fat.

We could go onto far more sophisticated measurements like Dual-Energy X-ray Absorptiometry (DXA) or Air Displacement Plethysmography, at which point things are starting to get expensive but the relevance for body fat measurement is increasing. Of course, our decision for what we measure will need to come with some experience and knowledge of how much resource we have. If we're planning a new product targeted at overweight people with the aim of keeping an eye on their progress, with hundreds of people to keep track of, we're likely to be restricted to the simplest, but not necessarily most relevant, techniques.

Even when measuring the right quantity, we can still get irrelevant measurements. My family have recently moved to a new house that has a thermostat which can be transported from room to room. This is very useful; if you are in a part of the house that's quite cold, you can simply take it with you and the heating will turn on to keep you at the right temperature. Unfortunately, I've found this system can be easily misled.

What I expect the manufacturers were intending is that the customers would place the thermostat somewhere in the room that would give a good estimate of the average temperature in that room, one that is relevant for the task of heating it. It would then be possible to reach the target temperature by increasing the time the central heating stays on, which should keep you comfortable and your fuel bills at a reasonable level. What my wife has realised is that by putting the thermostat on the windowsill, the measurement of the temperature becomes much lower than the average temperature of the room – a temperature that is completely irrelevant for the job of the central heating. Using this method, you can cause the heating to stay on permanently and make the house very hot indeed, whilst claiming that the temperature is set to the same target as it has always been...

Systematically Random

Once we have established that we have relevant measurements, we can now start to consider whether they give us good estimates of reality. Measurements can be flawed in many ways. These flaws lead to errors in our data that make interpreting reality more difficult in the best-case scenario, and could send us in completely the wrong direction in the worst. Under this heading, we'll discuss two major types of error that we should keep an eye on when hoping to obtain reliable measurements from our environment.

There will be occasions where something is overlooked in our measurement technique that causes the same type of error in every result we get; our tape measure may have stretched without us realising, or the scale could be misaligned on our mercury thermometer. Maybe the floor underneath our bathroom scales is not flat, meaning we overestimate our weight by about the same proportion each time. These are our 'systematic' errors.

These errors will typically influence the 'accuracy' of the measurements we're taking, and there will be things that we can do that make them better or worse. The estimate of car speed from a speedometer is heavily reliant on knowing the diameter of the wheels, as the measurement it takes is based on their rotational speed (this is much easier to measure than the longitudinal speed of the car over the ground). We can do things that will vary the diameter of the wheels that the speedometer will not account for, for example, we can fill the car with people and luggage and the distance travelled for each rotation of the wheels will reduce slightly. We can change the pressure in our tyres, or even change the design of tyre altogether.

These things will make the estimate of our speed that comes through to the driver slightly less accurate; in fact, car manufacturers may choose to deliberately reduce the readout of the speedometer just in case somebody puts the car in such a strange condition that the readout appears significantly lower than the speed the car is travelling. Customers will be very unhappy to be pulled over when the speedometer suggested they were obeying the speed limit! We could kid ourselves that we can get around problems like this by using more 'precise' measurements. Don't be fooled into thinking precise measurements are accurate. Just because you have a digital speedometer readout on your new car that reports to one decimal place, do not assume your estimate of car speed is more accurate than an analogue dial.

Even when we've done our best to ensure our measurements are accurate and free from major systematic errors, we can still find that they suffer from 'Random' errors, or 'noise'. By this we mean that while the average value of the measurements is close to the true average of the signal, it is subject to fluctuations at each point. This makes them 'unreliable' and can cloud the signal you're trying to isolate in a fog of randomness.

These fluctuations may have a cause; your electrical sensor might be receiving interference from another source, your survey results might be sensitive to the moods of the participants and times of day you perform them, your tape measure might bend when measuring over a large distance making it difficult to get an accurate reading. Unfortunately, we probably won't be able to isolate the cause well enough to understand it, and even if we can, there might not be anything we can do about it.

If these fluctuations are on the same scale as the change in signal we are measuring, we will find it incredibly difficult to determine the truth from the noise. The presence of these errors is not necessarily a reason to give up, as we will find there can be reasonably effective means of removing it; however this will usually involve taking more measurements than we'd ideally like.

Noise is a constant nuisance when interpreting the data we get back from the track, and lot of effort is put into reducing its magnitude to allow the true signal to come through as cleanly as possible.

Sensors like accelerometers are particularly bad for this, making it incredibly difficult to know how good or bad the performance of the car actually is without the driver being able to describe it directly.

We will often encounter noise in surveys like opinion polls prior to elections. A good poll will typically present a margin of error alongside the result, which will prevent us from drawing too many conclusions into a candidate's apparent 0.1% lead. These polls will typically vary day to day but should, over time, even out to capture the mood of the electorate reasonably well.

Noise will cause us big problems if we expect to use the measurements we take in calculations; if the magnitude of the noise is similar to that of our 'true' signal, our calculations could end up being some way off the answer we would expect. While there may be methods we can use to remove noise from our measurements, we should not expect to be able to completely overcome an inadequate measurement technique.

Don't Mind Me, I'm Just Observing

Imagine you enjoyed singing to yourself when you were in the house by yourself. After doing it for some time you might even find that you're getting better at it, hitting all the right notes. Now imagine that someone comes to your house saying they're doing an experiment on how people behave in their homes when they're alone, then set up some recording equipment and say they'll be back at the end of the day to review what they've recorded. Would you feel like singing? All of a sudden, maybe you don't act in the way you normally would.

Even when we can get our hands on the most accurate, noiseless measuring devices possible, we could still come up against this problem. There are some situations where the very presence of measurement devices might end up affecting the results you collect. This is known in science as the 'Observer Effect'.

In my own career, I've come across several situations where this can be a real issue. You will often see cars fitted with aerodynamic 'rakes' during testing and practice sessions, their job being to measure the air pressures (and therefore speeds) at each location using a sensor called a 'pitot tube'. These sensors are simple enough to build and use, and they will do a good job of describing the air flow around the car. But there's a problem.

Figure 16. Aerodynamic 'rake' on Formula 1 car

With the best design in the world, they will still cause significant disturbances to the flow of air towards the back of the car. They are also reasonably heavy and quite fragile. While this is not necessarily a problem for the points they are measuring, it means that the driver is very unlikely to be able to drive the car in the manner they normally would, which has consequences for the relevance of the data collected. While you may find the air behaving as you expect in the conditions you measured with, there is no guarantee that when the driver is pushing a bit harder, everything will remain as you think. The most annoying thing about the Observer Effect is that you will never know.

Most of the examples we've discussed so far have been reasonably formal and lead to a quantitative approximation for the part of the environment we're studying. We can however use the term 'measurement' for something far less formal. For example, If we want to understand how a friend is feeling on a given day, we can simply ask; we'll then interpret their response in the same way as we have all our other measurement examples. What they feed back to you could be inaccurate; they may be feeling better or worse than they are letting on. It may be 'unreliable'; perhaps just a vague description of what's in their minds. Or, it could be subject to the Observer Effect, where the very act of asking is enough to affect their answer. This kind of enquiry is probably more common than any of our formal measurements and might tell us more about the state of our environment than our sensors can. What I hope is clear from the discussion above is that we cannot assume they give us an accurate description of reality; the types of errors described will be true of most, if not all types of measurement. Their reliability will depend on the sensors we are using, how we're using them and the number of readings we're taking on each sample.

We should ensure we understand the uncertainty in the measurements we're taking before using them to make inferences. We can try measuring things we already know the values for, which will be how most sensors are calibrated in the first place. We can try using different sensors (both models and physical sensors) or surveys (if we're interested in people) to decide which is giving us most consistency. We can also review results to check whether the level of noise is as we would expect, given our understandings of the systems we are measuring.

We must be clear that we will be using the measurements we take of the environment to feed into how we should behave in future, which makes the understanding of their limitations all the more important. If we feel frustrated that they suffer from some of the issues above, there may be things we can do to help…

Noise Cancelling

Look on the pit-wall of any F1 practice session and you'll probably see a lot of people in team-branded clothing looking at screens covered in squiggly lines. The lines they're looking at will probably show readings from different sensors on the car, played out over time. The sensors will be measuring things like oil temperature, pedal positions and forces on certain components. Unfortunately, the sensor readings tend not to be much use on their own; it's not particularly interesting to know the rotational speed of a wheel without understanding how it's interacting with other areas of the car. What the people in the team-branded clothing will be there to do is put these individual readings together into something that gives them a better indication of how the car is behaving. They will be 'processing' the data.

The need for processing is simple; the raw information we recover from our measurement techniques might not tell us what we need to know about the environment. This could be because we aren't able to measure the quantities we're most interested in directly, or because the techniques we're using suffer from the issues described under the previous heading. Here, we will cover some basic methods for processing of data, as well as pointing the interested reader towards different areas for their research.

Figure 17. Formula 1 pit wall

One of the most common types of data you will come across is so-called 'time-series' (or time-history) data, which are simply measurements plotted out against the time that they were recorded. The squiggly lines on the monitor of our motorsport engineer are a good example of this. When it comes to the processing of time-series data, one of the simplest things we could choose to do is filter it. This is useful when the noise and the reality you're measuring are independent in the 'Frequency Domain'. As an example, anyone who's been to the seaside will know that the motion of the tides is very slow, with the sea rising and falling around twice per day. The motion of the waves, on the other hand, is comparatively fast. If we wanted to measure how the sea is affected by the tides, we could remove the higher frequency 'wave' motion with a filter and we should just be left with the low frequency 'tide' motion.

In practice, after removing our high frequency wave motion, we may find that there is some other 'content' in our tide-region of the frequency domain; any weather systems that might be passing through can affect the height of the water by lowering or raising the air pressure above it. This is going to make our lives difficult as we can't justify using a frequency-based filter to remove its effects, as doing so will remove some of our 'good' tide data as well. One thing we might be able to do is to take repeated measurements over a series of days or weeks - we should then be able to extract the repeatable motion of the tides from the less repeatable effects of the weather. This is an example of using our knowledge of the system in our processing.

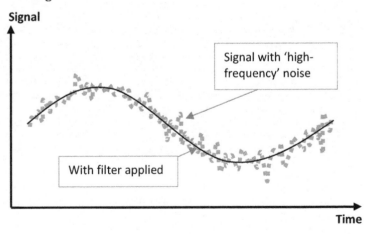

Figure 18. Example of filtering data to remove noise

One of our children's favourite activities is measuring their heights to see if they have grown any taller. They're often extremely disappointed by how long it seems to take to gain even a centimetre. Of course, the time it takes for a baby to grow to the height of an adult occurs over the timescale of years, so our children are going to have to put up with a centimetre or so every few months. Should we want to publish the huge amount of data we've collected in this study, we could again do some processing. If we were to measure their heights monthly, we are likely to be affected by sources of noise, such as any small changes in height that might occur on a daily basis, like the length of joints expanding or contracting slightly throughout the day. Provided we can take measurements frequently enough, we should be able to create a filter that removes the noise over the timescales of days/weeks and leave only the good data. This of course requires that there are no systematic errors in the way we are measuring.

We can use this example to demonstrate the problem of not measuring frequently enough, as not taking measurements at a high enough 'frequency' will mean that our noise becomes conflated with the signal in the frequency domain. If we only take measurements every year, we'll be unable to separate out any daily/weekly variation from the 'good' data that charts the change in height over years. This is called 'aliasing', and while the filter we use will make this data appear smoother, we shouldn't take that as an indication that it's more accurate; a mistake is very common when making inferences using filtered data. Again, your rule of thumb should be to only use filters that will allow any 'clean' signal to pass through virtually unchanged.

In these examples, we've been interested in measurements that have been 'polluted' in some way by sources of noise. We can now switch to discussing scenarios where we aren't able to measure what we're truly interested in but maybe we *can* measure other things that can be used to estimate it. For this we'll need a model that links our measurements to our interesting parameters, which we will call an 'observer'. This term is one we'll borrow from control theory and it describes something that can run in real time, or after the event, to extract more interesting results to help us in our decision making.

We can refer again to our racing example where we are being caught by a competitor car. The competitor car is not something we're going to be able to put our own sensors on, so we're stuck with estimating its state with information we do have. This may be the sector times and speed trap speeds of the car in question, as these are available to all teams. When we discussed this last, we explained that our course of action was going to depend on several key pieces of information: was the driver behind pushing their tyres too hard? Were they in a high-power engine mode that could only be used for a few laps? This is a useful case for an observer.

We should have reasonable models for the pace we would expect the following car to go at different points in the race. If we find a competitor is going much faster than we expect, we can start to make some assumptions about what might have changed. Our observer could help to estimate the state of the tyres, the engine modes, the effort of the driver, and give us a prediction of what we expect to happen in the future. Clearly, we have no new measurements in this process, we are simply using them intelligently to estimate things we're interested in, rather than just what we can measure.

There may be times when we are able to add or remove sensors at will, but these sensors aren't perfect for understanding what we're really interested in. For this we can turn to another observation technique called 'sensor fusion'. This works by taking whatever helpful measurements we can and using a model to fill in the gaps for us. While this is mainly used in an engineering context, this idea can be used to solve problems in other fields as well; for example, measuring employee job satisfaction in the workplace can be exceedingly difficult. Different people fill in questionnaires in different ways; questions can be interpreted differently, and your employees can have different ideas of what the 'strongly agree' and 'strongly disagree' mean. We can, however, measure things like employee overtime, sickness leave and attendance at work social events easily, which may be reasonable indicators of overall satisfaction, and a weighted sum of these individual pieces of information could be used to make a 'satisfaction index'. We can then make inferences on this value to determine any next steps.

Hopefully this is enough to demonstrate some of the scale of the subject of measurement processing. I expect that whatever the issue you're having with your measurements, you'll be able to find literature that describes how somebody else has approached the problem. The interested reader will no doubt find hundreds of sources describing tools with exciting-sounding names like the 'Kalman Filter', 'Auto-regressive Models' or the 'Hilbert Transform', each of which will have their uses and their assumptions. Your job will be knowing that they exist when presented with an opportunity that they're ideally suited for.

24

So, What Are You Going to Do About It?

Once we have our targets (dictated by our Utility Function) and our environment state estimate (that we've inferred from our measurements), we need to decide how we are going to act. As noted in our introduction, we can start with something as simple as a qualitative 'rule' that describes actions, or we can do something a bit more sophisticated. If we have 'measured' that we have run out of bread at home and what we really desire is a sandwich, we can go to the shops to buy bread, or alternatively find something else we'd like to eat that we have all the ingredients for. However, if we're reacting to stock market movements, we can decide how we're going to behave based on some mathematical manipulation of the measurements we are receiving from the market and the targets we have, with the objective of maximising the return on our investment. Similarly, if we desire a new product to be released to the market and we've measured that this is an appropriate time, we can put in place all our normal processes and checks, making sure these steps are followed through correctly until the product is ready for sale.

The control law will describe how we react to *every* possible output from the environment. We should find it simple to follow when everything is running smoothly but we will need to have something in place for when the environment decides it's going to make our lives difficult, which will include things like contingency plans and tests for whether to continue pursuing our original objectives. These will reflect what we would do automatically when faced with these situations; if we arrive at the shop to find it is closed for some reason, we'll probably try to find another one that might sell whatever it is we want to buy.

We'll discuss the types of control actions available to us in the sections below, starting with simple, discrete control and then move onto continuous 'regulator' problems. We'll also look at combinations of different methods before finally thinking about the action of experimentation.

Something we should touch on here is the lack of any process of inference or learning between the measurements we take and the control actions we decide to pursue. This is because the control process is purely mechanical; when the air conditioning in the car is targeting a particular temperature, it has no concept of how 'acceptable' that target temperature might be. Its job is to achieve that target with the ability it has available, and we should approach our control process in the same way. The process of learning about our preferences will sit outside this control; we'll leave this optimisation exclusively for when we make improvements in our process.

Be Discrete

We can describe some of our basic reflexes as control actions - when we have an itch, we scratch it. When we hear a noise, we turn to see where it's coming from. If we feel we are losing our balance, we try to steady ourselves. A lot of the time, we might not even realise we are doing them. These 'If-Then' type rules are very important in dictating how we, and other systems in our lives, behave, and are simple in as much as they can be described in a 'discrete' manner. By this we mean each process can be encapsulated in a series of individual operations. This is the opposite of a continuous process where steps are described by a set of laws that act on continuous quantities, like time, temperature, or speed.

While we will describe these as 'simple' rules, in reality they can be anything but. In the situation where we're losing our balance, we can be calling on many different muscles and sensors to do their jobs in a very short period of time. All we are doing by treating them as discrete steps is setting up some causality that describes how we are going to behave. We *measure* something like losing our balance, then *act* by steadying ourselves. We needn't worry ourselves with all the micro, unconscious processes that form part of the action, and should use descriptions that are tailored to the level of the hierarchy at which we normally operate.

Each of these steps may affect different parts of the task, and therefore our model. Most will involve making inputs to the environment, while some will involve measurements on the environment's outputs as well. There will be steps that require some of our resource to complete and others that involve systems where we are constrained by how we can behave. Common to all these steps, however, is that everything will progress in a discrete way.

Let's say we're about to launch a new product onto the market. If we've been through this process before, we know how we would expect to allocate our resources, the steps we must go through and the checks we need to put in place to verify everything is working as it should. We will also have contingency measures put in place if things start to go wrong. In reality, this is just one layer of complexity above our simple examples at the start of the chapter, as it is just a series of 'If-Then'-type actions strung together with some logic over which route to take. These processes can then be described using tools like flow-charts; each element of which will describe how the system behaves when activated. There might be a number of different routes we can take but these steps should adequately describe the control law for the task.

You may already be familiar with some of the other tools available to lay out this approach to our control process. Tools like 'Gantt' charts or Critical Path Analysis can also be used to tell us when we expect to perform each part of the operation, and which parts are dependent on others. Once we reach the end of the plan, we would expect to know how successful the outcome is going to be.

Many of the tasks I come across in my daily work will run like this, and can usually be distilled into a series of constituent elements that can be arranged on a plan like those mentioned above. The path we travel through the process will depend on the results we get at each step, until we arrive at our outcome.

When developing a new system for the car, we'll have a target in mind which is the characteristic we want to create. We'll then come up with a series of alternative designs to help achieve this target and evaluate these options with a program of offline simulations, simulator testing, lab testing and track testing, with different alternatives being dropped off the list as they perform worse in different tests of performance. The plan will exist as a critical path analysis with the delivery of the completed system to the track as an end date. By the end of the process, we should hopefully arrive at a system that will make the car faster in plenty of time to be useful to our final championship position (but this is never guaranteed).

Let's look at an example of a complete control process of this form, using the very simple example for making a cup of tea. We can build a flowchart for all the steps, including contingencies for if we do not have all the ingredients we need. The flow chart we can develop is shown in Figure 19, which considers both when we have all we need, and when we don't. In each path, our actions are based on the 'measurements' we get back from the environment.

As well as describing the process, we can use each step to estimate the expected utility. We can put cost and gains for utility on each part of this process. It is a negative when money needs to be spent, but this can be spread over many future cups of tea. Once we have what we wanted we should be able to confirm an increase in our utility, justifying the process we used. It may be that we won't achieve our desired objective, but this doesn't mean it was possible to do better.

Do It Continuously

We'll leave the simpler, discrete laws here for the time being. The rest of this section will explain some of the more mathematical ways we can perform the control tasks. (We'll also leave derivations and detailed formulae to other references - as with all things in this book, the idea is to convey the ways of thinking about these kinds of problems, rather than all the mathematics that go along with them).

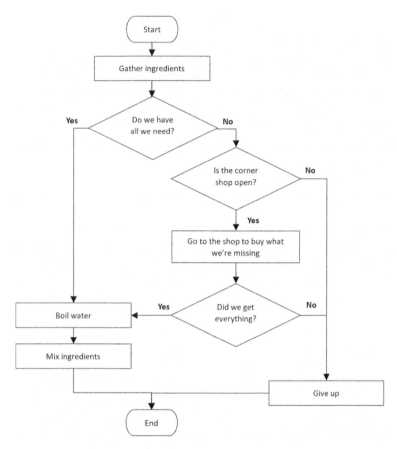

Figure 19. Simple control process - making a cup of tea

We will mainly focus on one common type of problem, which we'll call the 'regulator' problem. In this type of problem, we are trying to control our state to a specific value in a closed-loop way, meaning we base the strength of our control actions on feedback from environmental measurements rather than replaying an 'open-loop' sequence of commands. A good example of this type of control system in use would be a climate control system in a new car. Here, we want to maintain the temperature of the car at a fixed level that we know is comfortable for those inside, and to do this we'll need a way of dealing with 'disturbances'. These are changes in the environment that make our current level of control action inappropriate - in this example this could be a change in temperature caused when some clouds break up, allowing the sun to shine directly onto the car. The 'level' of air conditioning we will need to keep the interior at a constant temperature will probably need to increase at this point.

A common type of control law that is used to deal with this problem is called 'PID' or Proportional-Integral-Derivative, so called because there are three parts that dictate the control action that is applied: the proportional element, the integral element, and derivative element. This simple controller can help to explain some of the major ideas in 'Control Theory'; the science of hitting our targets. We will take a little time to describe how this works in practice, as it should help to illustrate different ways we could behave when dealing with these problems.

First let's consider the proportional element of our controller, which changes the 'effort' our climate control system makes using a multiple of the difference between the current temperature and our target. Think of this as like pulling on a spring; as you stretch it further and further from its original length, the force it generates to pull you back increases. The usefulness of this kind of control is reasonably obvious - if there's no difference between the current state and our target, we shouldn't make any effort to correct it, but as the actual state starts to move away, we should increase our effort in line with the distance to the target. As the temperature inside the car rises above the target through disturbances from the environment, the controller starts to act, giving more power to the air conditioning system to push the temperature back down again.

Something we will probably find when using proportional-only control is that, rather than settling at our target value, we'll reach a value that's close, but not perfect. We can explain why using the spring analogy we used before. Imagine one end of the spring is fixed and the temperature outside the car is like the force pulling on the other end. The target we want is to reach is the original, relaxed length. If the temperature outside the car increases away from our target, the force on the spring increases and it lengthens until its own force balances the one being applied. This is much better than having no spring (controller) at all, as we can still say somewhere near the original position (taking away the spring entirely means any force would be entirely unopposed), but we're never going to get back to the original length unless the force from the outside goes away completely. Instead, we'll reach a point where the air conditioning is applying a small, constant amount of effort that's enough to keep the temperature only slightly above the target.

One thing we can do to get closer without introducing a more complicated controller is to increase the stiffness of the spring (the amount of force needed to lengthen it by a certain amount). In our climate control system, this is called increasing the 'gain', or the amount of effort we will put into the air-conditioning and heating systems for every degree we go away from the target. This sounds like a good idea if it will mean we always end up closer to our target temperature, but we must use caution, as when we do this we open ourselves up to the risk of instabilities as the temperature 'overshoots' the target. This could happen if there are delays in the system, like the action of the air-conditioning taking a while to reach the temperature measurement sensor. The last thing we want is the heating and air-conditioning wasting energy fighting against each other, instead of the disturbances from the environment.

Rather than simply increasing the gain of the proportional controller, we can consider introducing an integral element. This doesn't only care about the *magnitude* of the error from the target, but also the *time* it has spent at this error. With this term, we can remove the constant offset from the target we achieved with the proportional-only controller, as now when the temperature in the car fails to converge on its target quickly, it gets an extra kick from the integral controller. We should now find that the temperature in the car settles at the target, rather than some intermediate level. Again, if we have delays in our system, we can find our integral controller 'winds-up', meaning it requests an increasingly large control action before it's had any effect on the system it's controlling. Again, higher gains will only improve the ability to control our system up to a point.

If we would like to reduce the magnitude of the overshoots we might see when controlling our system, we can introduce a derivative element to the controller. Again, this element doesn't measure the magnitude of the error, but rather the *rate* at which the error is changing. As a result, a derivative-only controller would not be able to target a fixed value of temperature and it should only be used in conjunction with at least one of the other two elements. This has slightly better predictive ability than the proportional-only controller, as a difference in the rate of change of error can be detected earlier than a change in the magnitude of error. This means controller effort can be applied sooner, with an effect of this being to reduce any overshoots when controlling back in the other direction. If not used carefully, this can still promote instabilities, as the controller can overcorrect small deviations away from the target.

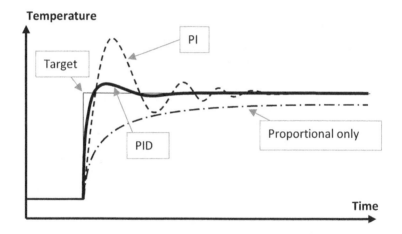

Figure 20. Illustration of simple controller types

An illustration of how the different controllers might react to a step change in their target is shown in Figure 20. Here we can see many of the ideas we've just talked about. The ideal proportional-only controller doesn't overshoot but the final value has an offset relative to the target. The PI (proportional and integral) controller doesn't have this final offset, but it overshoots the target and oscillates either side for a short period, before settling. The complete PID controller reacts earlier than either alternative with no final error. The overshoots are 'damped' relative to the PI controller. Hopefully you can begin to appreciate why this design is so popular; although, there is likely to be some tuning involved to get the best out of any of these options to get the response we're after.

Something that is key when dealing with these systems is to design controllers with a suitable length scale in time. If we have a system with some delays, where control actions take a while to feed through into the environment being measured, it is not a good idea to have a controller that can build up too close to maximum effort before it's even started to have an effect. As described, this will lead to instabilities that mean we completely lose control. Think of it as e-mail Vs conventional post; if you send a letter via the post office in the morning, it's no good getting upset if you haven't had a response in the afternoon and sending another letter to complain. We must be mindful of the time taken to see a response in the system before making the next move. This is less true with e-mail, which should be delivered almost instantaneously, and therefore if we do not get the expected response, we have more reason to be grumpy. By increasing the controller 'gains' beyond the point of stability, we will find ourselves losing control, rather than gaining it.

While this kind of controller clearly has its place in engineering systems, we can think about applying these lessons to other, less obvious situations that we may come across. Let's say you're tracking staff satisfaction through surveys and other measured data every month or so. Your Utility Function for staff satisfaction starts to roll off after a score of 80% (in arbitrary units), so there is little value in trying to push staff satisfaction over this amount. You can influence the satisfaction of your staff by investing more or less heavily in you HR department; giving them more to spend on employee assistance programs, social activities, more detailed investigations into low morale and so on.

Let's say after a few months you've noticed your scores start to drop to around 40%. Maybe there have been a few instances where employees have had to be pulled up for poor performance or have left to follow a different career path. These represent 'disturbances' to the system that are beyond our control. This fall in satisfaction is significant enough that you think it's worth some investment to bring it back towards your target value. How could you respond? One option is to mimic this type of PID control.

Initially we could start taking some action and allocating more of our resource to HR by taking it away from a different area; maybe one that our Utility Function is less sensitive to. The amount of investment could be proportional to the difference to our target (in this case 40%). Maybe after a few weeks we start to see some improvement (and we are confident that this is as a result of our investment) so we can start to reduce the extra amounts we are investing in HR and move it back to other areas. However, now it looks as though we settle at a new level, slightly lower than our target value, say 70%. We decide that we need some integral action to give a final kick up to where we want to be, so we increase the investment proportional to the amount of time we have spent below our target. Finally, to reduce our risk of overshooting, we start reducing the investment before reaching our target, as we know there are some lags in the system that will probably mean we reach the target before we are able to measure it. This is mimicking the derivative controller aspect.

The above example is some way from the idealised controller that might exist in your car's climate control system but hopefully it helps to show that some of the ideas used formally in the development of machines can be transferred to more qualitative actions.

While the above mechanisms should get you most of the way to our target, we may be able to do even better. If we understand the systems that we're dealing with very well, we can start to include this information into the controller we design, rather than treating them like a 'black box'. This leads to tighter control that is more aligned with our requirements. There are two kinds of methods that we will touch upon:

- Optimal Control
- Model Predictive Control

A major difference between these and a classical 'PID'-style of controller is that we use a model of our system in the derivation of the gains, rather than leaving this as a 'tuning problem'. With this approach, we understand the trades between the input and output signals before we've even started to use the controller. This will not be true in our PID case. Going back to our HR investment example, do we really know what we would need to pay the department to get a 1% change in the level of job satisfaction? If it's on the scale of £100, rather than the scale of £10k, we will find ourselves overshooting considerably when applying the corrective action.

The only tuning parameter we are left with in our optimal controller is a weighting between how tightly we want to stick to our targets and how much effort we are prepared to exert with our controller. In our climate control example, we probably don't mind the temperature drifting one degree or so away from its target if it means we can save some money (fuel) by doing so. This is a far nicer way of dealing with the problem than having to tune individual controller gains, though we will need a reasonably detailed model of the system before we can get started.

Model-predictive control takes this one step further. It opens up the opportunity to 'look-ahead' to upcoming disturbances (that we know about in advance) and take action before they happen. This is because it contains possible future inputs and system behaviour in its state, so the controller is not completely 'reactive'. This can be a more intuitive way to think of the process of applying the controller. Imagine trying to design a controller for driving a car that couldn't look ahead down the road; you would only be aware that you'd hit another car as it started happening. This is probably a little too late to start applying the brake pedal.

I've always found the lessons covered under 'Control Theory' immensely satisfying in their application; watching as your linear motor positions itself to within a micrometer of where you asked it to go gives a great sense of achievement. My time spent developing a simulator motion platform was full of these kinds of moments. Of course, there are mistakes, when something is fired at full speed towards the end-stops, but even these give you something to talk about. At the end of the project we had an extremely impressive machine that a person could sit in and get all the sensations they need to drive the car at the limit of grip.

This project also gave me a bit of inspiration to pursue some robotics projects in my free time. It's a hobby that I can definitely recommend to anyone who wants to polish their skills in control theory, electronics, programming or engineering in general. The great thing about it is the hardware can be bought for next to nothing - the microcontroller I have cost around £30 and with a bit of extra spend on some electronic components and motors (you'll struggle to buy enough to satisfy the minimum order on the website with these!), you can have enough to watch your newly created robot navigate around obstacles on your kitchen floor.

My only advice is to steer clear of mains electricity until you're confident you know what you're doing. Home insurance claims can be awkward enough at the best of times...

Combinations of Control Actions

If the tasks we're dealing with are complicated, with many different processes, we'll probably find ourselves dealing with both discrete steps and continuous regulator problems at the same time. We have already discussed the fact that 'simple', discrete operations can cover groups of individual operations (see our example of making a cup of tea example from earlier) but they can also contain regulator problems.

We can look back at our example of driving to work again here. The control law that describes the entire process might follow a flow chart-like structure, like that shown in Figure 21. We know the route we want to follow and what alternatives we can take if the road conditions aren't quite as we'd expect. This flow chart captures these steps but buried within them are some reasonably complicated regulator problems. These relate to our control of the car.

Driving along an empty road at a constant speed is not something that will happen on its own. To achieve this, we are going to have to control the car's trajectory to stay roughly in the middle of our lane with the steering wheel, as well as using the throttle and brake pedals to control the car's speed. If we encounter disturbances like hills or sharp bends, we are going to have to use these controls differently to when we are on a straight, level road. Parking is likely to require similarly complex control actions.

These regulator problems do not even consider the actions of other road users or changes in the road network. On our journey, we are also going to have to maintain a safe distance to other vehicles and react to unforeseen events while reducing the risk of a serious accident. The complexity of this control problem explains why we still rely on human beings to drive cars, rather than computers and algorithms.

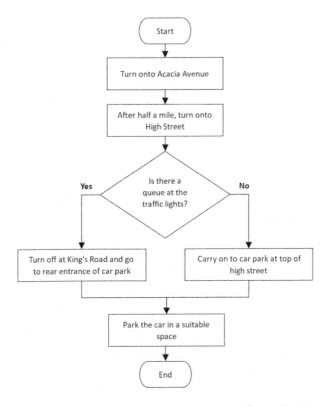

Figure 21. Example of a control process for driving to work

All this complexity doesn't stop us wrapping these processes into single blocks in our chart; most drivers will perform these steps automatically, without need for any serious conscious thought. It's only when something unexpected happens that we might need to turn the stereo off and focus on the processes that we normally perform without thinking.

Breaking down our tasks into discrete and continuous steps will help us in making improvements later. Our regulator problems are likely to require a little more finesse than our discrete steps, which might respond to 'brute-force' approaches. Reducing the time it takes to chop up firewood can be achieved with more powerful tools; reducing the time it takes to create a piece of furniture, on the other hand, will mean improving our precision. This is likely to involve more accurate measurements, better tools and greater care.

In this chapter, we have covered the different types of operation that we might use to maximise our expected utility in the control process. We have discussed continuous and discrete operations, as well as combinations of the two. These elements will be joined together and can be expressed using familiar planning tools.

We can think of discrete control operations as steps in the process that we'll need to execute to complete the task, and which can be arranged into flow charts that allow the agent to pick a route between their current state and the end of the task. The route they follow will be a function of their preferences and the feedback they measure from the environment. While we can describe these steps using simple language, they are likely to involve a number of smaller steps that exist at lower levels of the system hierarchy. We will probably find it most efficient to label steps using descriptions from the level of the hierarchy we're operating on.

Continuous operations, like regulator problems, will be sensitive to our inputs at any point in time. These are operations that will require us to control an element of the environment towards a specified target, and which will increase our utility in some way. These problems are likely to require some finesse and probably won't react well to big changes in their inputs, therefore our actions will require continuous monitoring of the outputs of the environment to decide what action to take next.

We will probably find that the control process we need to work through is built up of both discrete and continuous operations, and these will often need to be executed simultaneously to achieve the results we're looking for. While these steps will have a direct influence on the outcome of the task, there is one other type of operation that we might decide to include in our process.

25

Doing Your Research

Over the course of our task, we may find that we find we lack knowledge in an area that we believe will be key to our success; all our existing documentation and other literature on the subject falls short and we are struggling to decide what to do next. What can we do? Well, we could decide to give up and try something we understand better, or we could embark on an 'experiment'.

When we think of experiments, we normally picture groups of scientists in white lab coats holding test tubes of brightly-coloured liquid. Our definition will be somewhat broader. We will consider an experiment to be a process where we make inputs to the environment and follow these up with a sequence of measurements, analysis and learning. This could well be a formal, academic investigation with certain standards and a process of peer review, or it could simply be trying to do something we do every day slightly differently.

In this chapter, we will think about how we can design our experiments to give us the information we are really looking for. We'll probably find that the success or failure of our experiment hinges on having a suitable plan for what we will test. When it comes to interpreting the results, we need to make 'inferences', which will be left to the following section when we describe the process of 'learning'.

Hypothesis Testing

Let's go back to our driving to work example. Imagine that we've been particularly organised one day and we've got into the car 10 minutes earlier than normal. Rather than just driving slowly the whole way there, we decide that we will try an experiment. In this case, we'll try a different route. Our navigation app suggests the route we're going to try is slightly slower, but we know that this doesn't account for things that we've noticed on a day-to-day basis. Maybe the bin collections are happening that day, or there will be more cars parked near the school we would normally drive past because of a scheduled event? We don't necessarily expect the new route to be significantly faster, we just want to check that we're not missing out on something important.

On the day in question, we arrive to work a minute or two earlier than we would have expected to when using our 'normal' route. Should we switch to going this alternative straight away? Probably not, as we're missing a 'control' experiment. That is, we don't know how long it would have taken going the normal route on this day. It may be that the traffic was lower everywhere for some reason. It's likely that the fastest journey we have ever done on the normal route is still faster than the time we just measured with this new alternative.

To find out for sure (because who would want to waste time taking an inefficient route to work every day?), we will need to repeat this process a number of times, comparing measured journey times for each route whenever we do the journey. Once we've gathered lots of results, it's likely that our noise factors will have averaged out. When we compare the average times of each route, one will come out as quicker than the other, at which point we can use all the data we've gathered to perform a statistical test on the hypothesis 'the new route is no faster than the old one' (we always test against a failure to demonstrate the status quo, rather than any assumption over the quality of the alternative). The result of this test will tell us whether there is enough evidence to reject this hypothesis and start using the new route.

'Statistical significance' is the key thing we will be looking for from our results – one measurement of each choice is not an experiment. We need to be thorough when updating our knowledge of the environment or process we are using, as this will have knock-on effects down the line. This means removing as much noise as we can from each result to get the best estimate of reality. Once again, the 'design' of our experiment will be key.

The 'Design of Experiments', or DoE, is a critically important and very well studied area of research. While the above example evolved organically from an opportunity where there was little to lose, our experimentation in the real world is likely to involve some resource cost and time to complete. For these reasons, we need to make sure they are as efficient as possible, whilst maximising the power the experiment has to tell us what we need to know.

We have already shown that any measurements we take as part of our experiment are likely to suffer from 'noise', a term used to describe variations in our data that come from sources we don't understand in the context of our model. In all our experiments, there will be things that introduce noise which we can do nothing about. For example, if we're trying to measure the success of an outdoor exercise program, the weather will change continuously over time, as will any participant's mental state or physical condition. We know that these factors will be having an influence on the measurements we collect and that they cannot be removed during the experiment, but we can at least try to isolate them from our results to let the signals shine through. This will involve either taking enough samples for the effects to average out, or removing them with a separate (validated) model for their effects on the results.

Reducing the influence of noise on our experiment will help us attain one of the two possible successful outcomes of an experiment. These are:

- We have seen that our pre-held beliefs are correct so we can continue with these
- We have seen that our pre-held beliefs are not correct, and we have to change them

Similarly, there are two ways in which our experiment can be misleading:

- We have determined there is enough evidence to discount our pre-existing beliefs but these beliefs actually are correct - Type 1 Error
- We have determined that there is not enough evidence to discount our pre-existing beliefs but these beliefs are wrong - Type 2 Error

The designs of our experiments must minimise the probability of getting either error in order to maximise the probability of getting successful results. To minimise our chances of a type 1 error, we need to reduce the threshold for statistical significance of our experiment. This is the probability that the results we have could have happened by chance and are therefore still within what's possible from our prior beliefs. To minimise the probability of getting a type 2 error, we need to increase our experiments 'power'. This is the authority the experiment has to give a clear result, given the level of noise inherent in the process being studied. We can achieve both of these things by firstly maximising the number of measurements we are taking, and secondly, increasing the magnitude of the effect we are studying.

When it comes to sample sizes, we will typically be constrained by budget and time. We should be aware that more samples will rarely lead to lower quality of results, and we should always try to increase the number of points we take. We must also be sure that the environment of the test is as relevant as possible, meaning the diversity in the experiment should reflect the purpose of the test; it's no good testing medication for the elderly on only young people, for example.

I have been involved in many experiments where the magnitude of the change we were testing has made difficult for the effects to come through. A typical case might be where the race car is not behaving as we expect in a particular condition. Let's say that for some reason, the driver is complaining that they are unable to go on the throttle on the exit of one of the corners because they think the rear axle has very little grip at this point. The model captures this weakness well and we're trying to find solutions to improve the performance in this area. There are two hypotheses for what could make a significant difference; firstly, that the suspension design is weak for this condition and could be modified to weight this area more heavily. Alternatively, it is suggested that the car doesn't produce enough 'downforce' in this condition and more would lead to improvements.

We decide to design an experiment to test whether either of these changes could be a possible remedy, performed in the simulator with a driver in the loop. So how do we modify the model to best understand which of these approaches is best? We could choose to make a small modification to the suspension and the car aerodynamics and see if the driver can pick this up. After all, we probably can't make huge changes to the car after it's been built and the authority we have to 'cure' this particular problem is likely to be quite small. Alternatively, we could make much bigger changes than would be possible to make in reality. The strength of this approach is that it's much easier for the driver to pick out the effects when they are large. This will probably lead to larger differences in the lap times we measure during the simulator test, increasing the signal relative to the noise level inherent in the testing method.

The second approach is the one we should take. The first adds an artificial constraint of making a 'realistic' change to the car. We are not limited to achievable changes in the simulator and the purpose of the experiment is to isolate solutions, rather than implement them. Once we have confirmed which approach is more likely to achieve the desired results, we can think about possible means for achieving it on the real car.

When performing tests of this kind we might limit the total number of samples we take because measurements can involve certain costs we'd like to minimise, if at all possible. One approach is to use a different type of experiment known as 'sequential analysis', in which each result (lap time) updates a test 'statistic' that can be used to determine whether the experiment should continue or not. This means we can stop once we've taken the minimum number of samples to achieve the results we want. After each measurement is taken, the statistic is updated and compared to two values; the first is the passing criteria and the second is the failure criteria. If the test statistic doesn't reach either of these thresholds, another sample is taken. This can be a helpful approach if there isn't much resource available to conduct the experiment.

Regression

All the examples under the previous heading focus on obtaining an answer to a yes/no-type question - is x better than y? Or is x a cause of a problem we are experiencing? And this is not the only kind of experiment we could find ourselves participating in. The second type we will discuss is called a 'regression'. This type of experiment is useful when trying to create a model of an effect or 'response' over a set of design variables or 'factors'. When we use EXCEL to plot a line of best fit through our scattered data, we are performing a regression. We expect to use the formula of the line of best fit to predict future results based on the values of the factors.

As with many topics in this book, regression is the subject of a great deal of academic literature. We'll discuss some important elements here, but I'd advise the interested reader to look around for other sources. Chris Mack's free online course 'From Data to Decisions' is a brilliant introduction to the more complex areas of the subject.

Again, the design of our experiment will be critical here; we need to ensure we have the correct design for the variables we want to understand and how we expect our responses to vary with them.

Let's take the most basic kind of regression, a linear sensitivity analysis. This is useful if we expect the relationship between the parameter we're investigating and our metric to be linear in the range we are studying. If we have a limited number of samples, regression theory suggests we maximise the size of the changes we apply.

For example, consider the case of an airline wanting to understand how plane weight affects fuel consumption, and therefore how much effort they should put into reducing the weight of their planes to save money. If we believe the relationship between weight and fuel consumption is linear (which it should be over relatively small changes in weight, that is +/- 10% rather than 10x the weight of the plane), the most efficient thing we can do is run all our experiments at either maximum plane weight or minimum plane weight. This gives us the best possible chance to overcome the noise that arises from factors like different weather conditions on each flight, differences in pilot styles, inaccuracies in the measurement of the aircraft weight etc. This is similar to the reasoning we used above when discussing results of a hypothesis test. This experiment design is known as the dumbbell design because results will look like a dumbbell when plotted on a scatter plot, with all points at one end of the x-axis or the other with noise to give them some height. Other designs with intermediate points are useful for understanding the 'curvature' on the response but will reduce the power of the experiment to find 'first-order' (linear) effects.

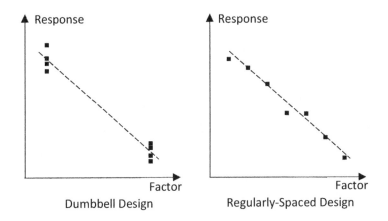

Figure 22. Illustration of dumbbell design and regularly-spaced designs

For more complex relationships, we will need more complex designs. Every time we add complexity to our model, be it with more factors or extra terms, we increase the number of samples we will need to take. Maybe we want to understand how a range of factors, such as age, height, weight, IQ, salary, qualifications, number of children and political views affect the results of our survey. Each sample (survey result) will have a different combination of these factors and we will need a lot if we want to understand how they all contribute to our results.

To reduce the size of our experiments, we should consider 'screening' our design variables to see if they really have an effect on our response, as it's quite likely that many of the factors you are considering don't have a large enough effect on the response to warrant any further investigation. It's no good designing a very sophisticated experiment, collecting a lot of data, only to find most of the factors you are testing have no influence on the responses you are measuring.

This is a common occurrence in my research – annoyingly there is seemingly no end to the number of ways you can make next to no difference to the performance of a Formula 1 car. This could be small changes to the suspension wishbone designs, different gear ratios or changes to dampers. Generally, the only means of making your way towards the front of the grid is with more power, better aerodynamics, treating the tyres well or getting a better driver.

Your screening experiment will be very simple and will only aim to understand whether you should carry each factor through for further investigation. Once you have a reduced set of factors (having removed some based on the results of your screening) you can design a more sophisticated experiment with which to populate your model. This is likely to be far less costly than your original design, even when accounting for the cost of the additional screening tests.

Again, the most efficient experiments will take only the minimum number of samples required to collect the necessary information. You should bear in mind whether it will be possible to collect more when you reach the end of the test, which can be a contingency if, at the end of the experiment, there is still too much noise to pick out the signal. Getting it right first time is always difficult, given that you don't know what the results are going to look like before you've performed any tests.

Possible Traps

If we're not careful with the design of our regression experiment or hypothesis test, we can end up misleading ourselves with the result. When experimenting with which route to take to get to work, a possible design would be to do route one every day in August and route two every day in September which should, after all, give us plenty of data to look at. There is a problem though. We will find there is probably very little traffic on our August days because there are no cars taking children to school, whereas there will be plenty of these on our journeys in September. What we have accidently done is 'confounded' a noise factor in with our results; so don't be surprised if on this occasion, route one looks like the best route to take every day.

Avoiding this kind of biasing is simple; we need a process of 'randomisation'. We could toss a coin before we start our journey to decide which route we take, which will do away with any 'unconscious' bias towards one route or another. Similar practices are performed in medical experiments when selecting which of the participants will be in the control group and which will receive the treatment. Not randomising in this way could lead to a higher probability of getting one result or another.

Of course, it is possible that someone conducting the experiment will like the idea of getting one result over the other very much. If you have been invested in a particular area of research for some time, not getting results you were hoping for can be very disheartening. This kind of 'intentional' biasing is particularly prevalent in industry, where people pedaling their own prejudices may seek to influence results (or at least ignore the results that contradict the previously held beliefs) to gain standing in a company. It's human nature to want our experiments to give interesting results but most of the time, the universe doesn't play ball. You're then left with a choice of facing reality and (perhaps wrongly) getting no credit, or trying to exaggerate the positive results you did get, whilst dismissing all the others. In these scenarios, we must remember to act like scientists, not lawyers. Our experiment should be designed to show the truth, not reinforce own interpretation of it.

We can prevent biasing by holding blind or double-blind studies, during which the experimenter and/or subject aren't aware which level they're testing. In simple experiments, this could be not knowing whether a participant is in the control group or the group that receives the 'treatment'. In sensitivity studies, they would not be aware what 'level' of design variable is being applied. An effective program should eliminate the possibility of biasing, provided there is no possibility to interfere with the results after they have been collected.

Taking another step up in our experiment's complexity, we can consider a relatively new field of study in experimentation; so-called 'Big Data', or data mining, which is the process of taking large amounts of data and picking out trends that we can use in our models. Again, this is a fascinating subject, full of interesting techniques, but we should be wary. The scientific method must begin with a hypothesis, and throwing a load of data into an algorithm and expecting it to come up with a perfect model to explain the results, based only on the trends seen in the data, cannot be considered scientific. What we are looking for is causality; that is, a change in the value of a factor has *caused* a change in the response we're measuring, rather than just correlation, where our response only changes at the same time as our factor. Here are some examples of correlations seen in mined data that you'll be able to find online*:

- Global temperatures correlate with a decline in the number of pirates in the Caribbean Sea
- The number of people dying from drowning in swimming pools is correlated with the number of films released starring Nicholas Cage in that year
- The number of sociology doctorates awarded is linked to the number of Worldwide non-commercial space launches

Instead of drawing conclusions too early, we should be using these results to create hypotheses to use in scientific experiments. Once we have seen there is a trend between the colour of the clothes in our display at the front of the shop and the amount of money people are spending inside, we should be able to design an experiment to confirm or refute this, rather than accepting it as true from a data mining technique alone. The key point here is that we cannot draw firm conclusions from data where we have no idea of the conditions that caused it.

---

*See https://www.tylervigen.com/spurious-correlations for more of these

Hopefully, by the end of the experiment, we'll possess a set of results that we have a high degree of confidence in. However, we need to be very clear over what kind of scope we think they are valid within. It's all very well discovering that we can increase the grades of children in a certain subject with a different teaching method, but don't expect it to improve the grades of all children over all subjects.

The purpose of our experiments is to gain knowledge in a particular area. Science is the search for what is true, and our experiments should bear that in mind. You will meet, or will have met, certain individuals who will claim to be able to extract 'knowledge' from incomplete experimental data that does not pass the above tests for robustness. One test that we should all have for knowledge is whether it would be accepted for publication in a scientific journal; if we have a well reported, double-blind study with high experimental power and low significance level, with a comprehensive list of the limitations of the findings, we probably have a good chance. If we have hearsay and speculation based on shaky data, you'll be laughed out of the building. The findings that are being pushed may well turn out to be true, particularly if supported by theory, but without passing the above tests, they cannot be thought of as knowledge. I wish anyone who disagrees good luck.

26

Putting It All Together

So far, we've covered the main ways we might act to complete the task we've set ourselves. There are steps that can help us progress towards completion and steps that can reduce our uncertainties and make the rest of the process more efficient. Hopefully we're now beginning to see what our control process will look like. In this chapter, we'll cover some final considerations that will be important for its design. These will be concepts that we should look out for, rather than any new ways we could choose to behave.

Firstly, we'll revisit how constraints will restrict our behaviour, before going on to discuss how the concept of state relates to our process. Again, we will need to understand how uncertainty will affect us before we look at an example of what a complete process might look like.

Constraints

In my current position, I am reasonably close to the beginning of the development process for new car parts or systems. On any day, my team will have, or might be given, an idea that is intended to improve the performance of the car, and our job will be to put it through its paces using various kinds of simulations. A lot of the time this doesn't really lead anywhere, with the new idea either proving no better than what we already have, or too complex to justify any resource spend. Occasionally, however, we will find something that looks as though it can make a significant difference. The theory, simulation results, and driver comments all point in the same direction; that this is something we should develop for the car straight away.

This is the kind of result you go to work for, and it can bring a great deal of satisfaction. Finding lap time through your own creativity and hard work is exactly why you wanted to work in this industry in the first place. You are often anxious to pass on the good news and set the design department to work on their new project. You will be asked to arrange a meeting where you will introduce the concept and discuss how is best to proceed. After finishing the presentation and clearing up any technical points that are unclear, there is always one question that will be asked; when do you think we can have it to run on the car? Unfortunately, the answer is always a disappointment. The idea has hit some constraints in our processes.

From my position, it seems outrageous that you can have a new system that will bring performance to the car which can't be available for several weeks or months. It may be that several races will pass by with the idea sat on the shelf, waiting for someone to pick it up and develop into a physical system for testing. We are all competitive individuals and seeing opportunities like this pass is frustrating, but that's only because we're not considering how the system is constrained. The design department aren't sat idle for the whole season, waiting for a new idea to come around; they are busy working on the current and following year's cars. Fitting in something new that isn't part of the original plan will be difficult, as it will mean pulling at least one person off their normal duties. The resource budget available is limited and they will already be performing other processes that take time to complete.

We can think of these as instances of the kind of constraint we discussed in our section on the task environment. Again, these constraints can come in several forms but can mainly be categorised as either natural or systems constraints.

To recap, natural constraints are things like the laws of nature and limitations of the natural materials and systems provided to us. Our systems constraints will be a product of the systems that exist around us, like the time it takes for your mail to be delivered; we may be able to influence these constraints if we throw enough money at them, but this will be undesirable in most circumstances.

Something we'll introduce here is the idea that we will be constrained by the resource available, be it monetary, time or skills. In the example above, we are unable to drop everything to work on a new design because there are other essential tasks going on simultaneously that require the resource we've been allocated. These tasks will also be monopolising the time of the design department who could be working on the new system you've conceived of. These are constraints imposed on us by the agent, and whatever the control process we design, we can only do it with what the agent has available to apply.

This shouldn't come as a big surprise. When we're deciding how to spend a weekend, constraints will limit the activities we have to choose from. While there are still plenty of things available to us - maybe make some home improvements or go shopping - we'll clearly be unable to take a holiday on the other side of the world as we only have two days. Buying a new house or car is also likely to be out of the question due to other resource limitations. Composing a symphony is likely to be outside the set of skills we possess at that time. It's not to say that we can't do any of these things at some point in our lives, given the time and money we would need, but these are not things that are likely to be achievable over a single weekend.

As we've already touched upon, these constraints may well have dependencies on things outside the task being considered; for example, we may have to avoid using certain machine tools at times when they have been allocated to other production tasks, or we may have to fit our review meetings in with the room booking system for the building. We might have to perform the task within times defined by every employee's contracts. We can think of these as a subset of our systems constraints, which relate to the systems that the agent is already part of.

In any case, these constraints are things we're going to have to keep in mind when designing our control process; it's not that what you want to achieve will be impossible, it will just have to navigate through these hurdles in order to become reality.

A State of Control

In the previous sections on the agent and the environment, we discussed the idea of their current 'state'; the set of quantities that describe the condition of the model element at a given point in time. While the control process is mostly concerned with changing the targets from the agent into actions on the environment, there may be internal states that we will need to keep an eye on.

The state could be used to describe where we find ourselves at any instant in time. The design of the new car system that we discussed earlier plays out over a reasonably consistent process. Anyone taking over the project halfway through would obviously like to know how much of it is already completed, and if we're just going through the final sign off on our completed design, it will be no good planning some brainstorming to come up with some initial ideas.

Part of this monitoring might include things like how people are allocated to different parts of the project. Individuals can move around depending on what is playing out at that time, so might be appropriate to include the distribution of work and skills as part of our controller 'state'.

A more mathematical example is integral element of the PID controller we discussed in an earlier chapter. While it should be possible to calculate the proportional and derivative elements of the controller without knowledge of the past, the integral element will need to retain a history of how long the current state has been away from the target state. Without this, it will simply reset at every time period and give no inputs to the process. Here, we will need to hold an internal state for this part of the controller.

Maintaining knowledge of the current state is essential for management of the task from start to finish, therefore as part of my management duties I have a reasonably detailed project planner. This lists all the projects we need to complete, together with their priority and those tasked with completing them. This gives a good level of detail on the 'state' of the team relative to our target of getting everything done. If a disturbance comes along, like an emergency project that requires immediate attention, understanding who is working on what makes it easier to move things around to accommodate any changes. This isn't intended to sound like a masterclass in management but should hopefully illustrate how monitoring the state of your control process could look in reality.

Uncertainty

As with all the areas we have discussed so far, the control process suffers from uncertainty. We've already covered some of the ways that this uncertainty can present itself when discussing the agent and environment parts of our model, which are still relevant here as they feed directly into our control process. To recap, from the agent we have uncertainty in our targets and our resource level, and we have uncertainty in the inputs we give to the environment and on the outputs that come from it.

Something we've introduced as part of the control process is uncertainty in our measurements. As discussed earlier in this section, these arise from the finite accuracy of the sensors we are using to provide us with information, which in turn, leads to uncertainty of our current state and therefore the error in our current state from our target (particularly given the target is uncertain as well).

Now we can consider uncertainty in our control law itself. This is the uncertainty in the way the process will play out, in absence of any change in the targets or disturbances from the environment. It is possible that in two different scenarios, the control actions will be different even if the inputs to the process look the same, which can be caused by elements of the control process being unavailable for some reason or affected in such a way that an alternative route is more efficient. We also cannot rule out that these differences may occur because there have been mistakes or oversights.

Let's take an example of a machining operation as part of a manufacturing process. The operations required on the component are always the same and we have produced many components according to this plan before. Let's imagine there is a cutting operation that uses a bespoke machine specific to this part. Unfortunately, this machine is being used as part of a different project and won't be available to us for another week. We could choose to wait for the machine to become available again, or we may decide that because this batch only has a small number of components, we can use an alternative machine to complete this stage. This would take slightly longer than using our preferred machine and might lead to some mistakes that mean replacements need making, but it will be much quicker than waiting. Which course is better for the agent who's in control? It will be up to the agent's Utility Function to understand.

For an example of how mistakes can affect the progress of our task, consider the last time you or a close friend/relative assembled some flat-pack furniture. Were all the steps implemented correctly at the first opportunity? If your experience is anything like that of most people, the answer is probably not. There are two possibilities when we make mistakes like this. We can either carry on and assume the mistake won't have a significant impact on the outcome, or we can go back and repeat the process in the hope that the same mistake won't be made again. There isn't necessarily a right or wrong answer here, it will be entirely situation dependent; if when we are constructing a table we forget to put in one of the five screws that holds one of the legs to the top surface, we may decide that it will be alright as it is. If we discover we haven't put in four of the five screws, we should probably consider how much deeper into the process we go before trying to address that.

It's possible that in part of the control process, two possible routes through the flow chart will be equivalent or give us virtually identical results. This is particularly true if we are lacking information about the consequences of our decisions. For example, imagine we're running a car washing operation and two customers have asked us to wash their cars before going off to do some shopping. Without knowing who is likely to come back first, we could come up with arguments for doing either one of them first. If we have two or more paths open to us and we're lacking information as to which would be best (without enough time or the ability to find out), any of the paths would be valid. We should be clear that this differs from uncertainty in our target, like in our 'running out of bread' example, we may choose to go to the shop if we suddenly decide it's very important that we have all the ingredients for a sandwich, or we could be very happy finding something else to eat. The correct choice will therefore be dictated by the Utility Function at that point in time, rather than by random chance.

Finally, we may have uncertainty because we haven't considered how we will respond in the event that we get a certain output from the environment. This is the control equivalent of being 'lost for words'; there will be times when something so out of the ordinary happens that we won't immediately know what to do. This is not necessarily bad planning, as we surely cannot account for every possible scenario when designing our process. There is, after all, only so much time in the day, and you will not need to look far back in history for the kind of events that might catch those involved completely unprepared. Heavy snowfall in the UK usually shows up inadequacies in the preparations made by those in charge. Flights are cancelled and roads are closed while other countries that experience this kind of weather more regularly basically carry on as normal. Despite these occasions, we should consider whether investing in preparations for future occurrences is really worth it, given the rarity of this kind of event. To invest in preparations that we never expect to use will be inefficient.

It will not only be the steps in the process which must be considered uncertain, but also the system's constraints. This will clearly not be the case for laws of nature, but will be true for things like the time it takes for parts to be delivered or the resource available for allocation. When we post a letter, we are aware that it won't reach its recipient on the same day, but can we guarantee that it will arrive within a week? If circumstances are against us, then no. Again, we must be sure that any process we are working on has enough slack to accommodate possible oversights in these elements. Expecting the project to take no longer than the best-case scenario for the critical path is going to lead to disappointment.

Example of Complete Process

With all these considerations in place, we can start to think about what a complete process might look like. The following is a relatively formal process, with plans put in place well in advance of the task being performed. Of course, the control process will not always look like this. There will be occasions where the task is so simple we don't need to write anything down. For others, the plan is something that will evolve over time, based on information we gather during the execution. We should still consider these as perfectly valid ways to behave, which are supported by our model, but they aren't as useful to us here.

One of the most impressive feats of organisation during a race weekend is the qualifying hour. This is the part of the weekend when the cars will be pushing for maximum performance, running as light as possible, with the engines turned up to give their peak power. This should enable the driver to set the fastest lap time they can in each segment that they're participating in. Getting a decent spot on the grid will mean starting ahead of your competitors, putting you in the best position to score points in the race. Achieving this will require maintaining the cars in their optimal running condition, keeping the driver in the right frame of mind and managing your performance relative to your competitors.

This will not only involve the driver in the car but also the team of engineers in the garage and on the pit wall. If you're listening to the team intercom during this time, you'll find that there's such a flurry of activity it's difficult to pick out what's going on. This is amplified further when trying to simultaneously follow the progress of both the cars being run by your team.

In this example, we will consider part of the qualifying session and come up with a simplified control process that describes how we should behave. We will need to use a lot of the techniques that we've described in this section, like discrete steps, regulators and constraints. The session can be broken down into separate sub-processes which need to be executed in turn to set a good time. I will use several different flow charts to illustrate each step; our aim being to beat our competitors to get as high up the grid as we can.

Let's imagine we are close to the end of the first session. In our previous run, we set a time that is currently fast enough to take us through to the next segment but there are cars behind us that are capable of going quicker and knocking us out. There is enough time left in the session for anyone to do one more run if they want to and, if we want to guarantee getting through, we are probably going to need to run again and go even faster. Of course, if our competitors are unable to set another time (problems with the car, they run out of new tyres etc.) we will not need to run, and we should stay in the garage.

The description above contains information that is relevant to the current state of the system. To understand how we should behave next, we need to understand the condition of the environment, the agent (us) and the control process. We can describe things like our current desires (to get high up the grid as possible) and the resource available (tyres, fuel etc.) as members of our agent's state. Ideas like the time left in the session, the weather conditions and the performance of our competitors can be considered as part of the environment state. Finally, any quantity that is required to drive the car, like the condition of the tyres and the driver's thought process, can be considered as members of the control process state. These are the things we need to keep an eye on if we want to get the best result we can; as always, we're not going to know all these things with certainty, so we'll need to choose the behaviors that give us a robust solution.

The first step in the process is waiting in the garage. For this step, the control process looks like the one shown in Figure 23. This is what we'll use to decide if we need to go out and do another lap. Firstly, we're going to have to keep the car in a state that means we're ready to go out as soon as we see our competitors doing so. This is a regulator problem and will require things like maintaining the engine at the right temperature before firing up, keeping the tyres heated and having the driver in the car ready to go. We could choose to take measurements of all these quantities as part of this step. If we didn't keep the car ready, we might save some money in electricity and the driver might be a bit less fatigued from having to stay 'in the zone', but we wouldn't have time to react to the sight of a competitor leaving the garage at the last minute. This would lead to being stuck in the garage while they did a lap that could be faster than our quickest, so would not be the most robust approach.

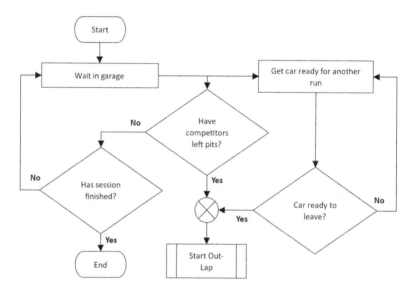

Figure 23. Waiting in garage process

Of course, we might reach the end of the session without anyone else going out, in which case we will go through without having to do another lap. We can stand down and get ready for the next part of the session.

In the overwhelmingly likely scenario that our competitors leave the pits for another run, we are going to send our car out as well. Now we need to get the car around one lap of the track at a reasonably slow speed before we can start a 'flying' lap. This is called the 'out-lap'. Again, there are going to be plenty of things to keep an eye on. Our flow chart for this part of the process is shown in Figure 24.

We have several regulator problems to deal with. Firstly, we need to get the car into a state promising the best performance when we get to the braking zone of turn one. This will mean keeping the engine, brakes and tyres at the right temperatures, while also make sure the driver is in the right frame of mind to deliver.

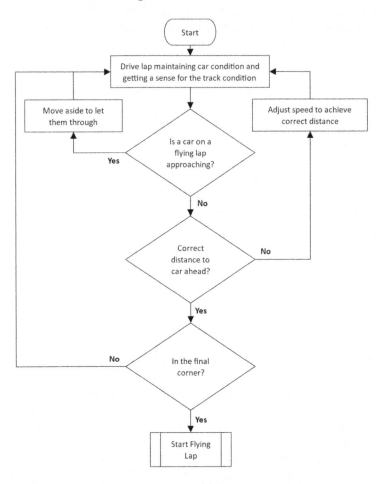

Figure 24. Out-lap process

As well as this, we need to manage our gap to the car ahead. Too close and they might get in our way on our quick lap, too far away and we risk being overtaken by the cars behind, making the job of maintaining the car temperature even more difficult. Driving too slowly will also risk the session finishing before we've had a chance to start our flying lap. We can think of the time left in the session as a hard constraint that we are going to have to deal with to avoid an undesirable result.

The out-lap will offer the driver an opportunity to learn more about the condition of the track before they start their quick lap, which could lead to doing some experiments. If the track is wet, the driver may choose to try out some different 'lines' to understand which parts of the track have the most grip. This is a common sight during a wet qualifying session, as the areas of track with most grip in the dry may have more water on them than less used parts of the track. Getting the most out of this opportunity might lead to some knowledge that other drivers don't have, offering a crucial advantage. Of course, we're not going to be able to be certain of these results given that we're only going to collect information from a single out-lap, but we're not trying to publish anything here; just looking for clues on how to drive.

During this lap, there could be events that will cause us problems. Any other cars that are on their flying lap will need to get past. Interfere with their lap by not getting out of the way quickly enough and there will be harsh penalties. This is an example of a possible disturbance from the environment that makes our task more difficult. If we see there is a risk of a fast car coming up behind us, we should make sure our driver is aware over the radio.

Once we've reached the end of our out-lap, the next phase of the process is really very simple indeed (see Figure 25).

Whilst this description is simple, this instruction involves a lot of effort; the driver is going to have to make sure they hit all their braking points, avoid spinning the wheels on corner exits and select the right gears for all the corners. By this point of the weekend, this should all come automatically; even relatively large corrections to the car trajectory will happen by instinct, with basically no thought required from the driver. The best laps will come when the driver and car are working in perfect harmony, with the driver able to react to every disturbance that the car or track throws at them. A quick YouTube search of some of the best qualifying laps in the history of the sport should hopefully bring this across.

Once we've finished our lap (and we're hopefully congratulating ourselves on progressing to the next part of the session), the job isn't over; now we must complete the 'in-lap'. Like the out-lap, this is a lap that we'll drive below maximum speed to return the car to the garage, where it can be prepared for the next run. Again, we're going to have to manage the car temperatures until we make it back to the pits to prevent any damage. Like the out-lap, we'll need to avoid any cars that are still on their flying laps - the flow chart for this part of the process is shown in Figure 26.

Figure 25. Flying-lap process

When we make it to the entrance of the pit-lane, we need to ensure the driver activates the pit-lane speed-limiter. The pit lane speed limit is another constraint that we're going to have to adhere to, since not doing so will probably lead to a fine that will affect our resource, and we might also get randomly selected by the stewards to be weighed. This is a process to ensure we're not purposefully running the car underweight to gain an advantage and have better chance of qualifying at the front of the grid. Again, this is a possible disturbance from the environment that we may have to deal with.

Now we're back in the garage and the session is over and if the process has gone well, we'll be through to the next segment. Unfortunately, even if you've executed the plan to perfection, you can still find yourself out and starting towards the back. I've been in this position plenty of times. Sometimes our competitors just have the faster car…

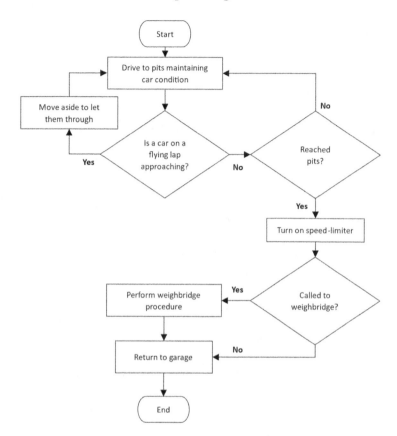

Figure 26. In-lap process

Hopefully you have seen examples of the ideas we've discussed in the section so far. This process involved both discrete steps and regulator problems. We needed to keep an eye on the state of the car and the environment and there were plenty of disturbances and constraints, which are likely to appear in any similar task we could consider. In reality, we're probably not going to have to lay our task out in this format to get the most out of it but hopefully we've demonstrated how this thought process can help you to identify the important considerations, and therefore the best way of treating each step to get the result you want.

27

Control Process Summary

In this section, we've discussed the different elements of the control process. The following is a brief summary of the key points we should take forward to the discussion of the final model.

The control process is what we'll use to determine what inputs we're going to make to the environment to achieve our objectives. The inputs to the control process are the target for our task and the outputs we receive from the environment. The target is dictated by the Utility Function of our agent, and we will use the outputs from the environment to make an estimate of its state.

Before we can make these estimates, we must perform measurements on the environment outputs, which can be formal (like sensor readings or survey results), or informal (like asking questions of others or perceiving through your senses). These measurements must be relevant to the area we are interested in and we shouldn't be distracted by what is easiest to measure. They may suffer from systematic errors, meaning they misrepresent the state of the environment, or random errors, making it is difficult to isolate the state of the environment from the distortion provided by the sensor. We should also be aware of the Observer Effect, which tells us that we may have altered the state of the environment simply by taking a measurement.

We may choose to process these measurements to increase their accuracy for the state we are interested in, using a technique like filtering. For this to improve accuracy, the noise must be distinct from the signal we are trying to measure in the 'frequency domain'. We can choose to use an observer to estimate the values of anything we are unable to measure directly, which requires a model to link possible measurements to the unknown values.

The Control Law is the process that takes our agent's targets and an estimate of the state of the environment and turns them into the way we're going to act. This can be split into either discrete or continuous actions but it's likely that a combination of both will be necessary to optimise the task utility. Discrete actions can be encapsulated by a flow-chart structure, with gates to decide which paths to take, while continuous actions are described by rules that judge what action to take at every point in time they are working.

We could also choose to perform experiments as part of the control process. These will not usually affect the outcome of the task directly but will help to reduce uncertainty. Our experiment design must help to maximise the probability of getting a successful outcome, which will require strong statistical significance (low probability of seeing the results from chance alone) and high power (the ability to pick out a trend from the noise associated with the process).

We have introduced two main types of experiment in this section: Hypothesis Testing and Regression. In Hypothesis Testing we are looking to understand whether making a change has influenced our results, while in Regression we want to understand how varying different quantities affect the measurements of interest. We must do all we can to avoid biasing in experimental studies, including double-blind testing. Outside of these techniques we could choose to 'mine' Big Data sets to look for trends we are interested in; however, we should only use this to create hypotheses for experiments. The process of mining data is not an experiment in itself because there is no understanding of the circumstances that brought about the results.

Our control process will need to respect constraints of the systems we're dealing with, which not only includes the types we introduced in our section on the environment, but the agent's finite resource as well. We'll find that the path we take through the control process is a function of these constraints and any uncertainty in the process. This uncertainty can arise from many different sources, in our targets and measurements for instance, but also because of things like mistakes or the fact we may be indifferent to different routes through our flow charts. It may be necessary to track the state of the control process if there are elements that do not fall under those of the agent or the environment.

So now we have a process to link our desires with the possibilities in our environment, and in a lot of cases we'll have all we need to produce a satisfactory outcome. However, if we want to extract all we can from a particular task, we're going to have to be intelligent about how we go about it. We will find out how in the next section.

Part Four

Learning and Optimisation

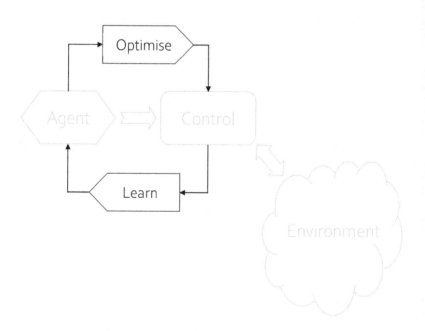

28

The Team is Dead, Long Live the Team

If you watch a Formula 1 race today, you might be fooled into thinking that each of their owners decided on one day that they were going to set up a team. They then put in some investment, hired the necessary personnel, bought a factory and then set about designing their first car. Once this was ready, they took it racing and have been doing so ever since. The reality is somewhat different.

Look back through the history of each of the ten teams and you will see that they have not always been like they are today. Let's take the example of the most successful team in recent memory, Mercedes. In reality, this team has little to do with the automotive manufacturer run out of Stuttgart in Germany. The team is based in Brackley, Northamptonshire (UK) and the operation was producing F1 cars a long time before Mercedes became involved in 2010. Prior to being bought by Mercedes, the team was called 'Brawn' for a single year in 2009 where it achieved the remarkable feat of winning both championships in their first season. Brawn had been born out of the remains of Honda, who had acquired the team in 2006, before pulling out during the economic downturn of 2008. Prior to this, it was the 'BAR' team. This team had been called 'Tyrrell' before that, starting from its founding in 1971 through to 1998. While plenty of people will have come and gone in this time, I expect there are employees who have stayed in the team through all these rebranding exercises.

This is not the only example of a team changing hands on multiple occasions. The Alpine team, based in Enstone, Oxfordshire, is remarkable in that it has been called Renault on *two* previous occasions, between 2002 and 2011, and between 2015 and 2020. In the intervening period, it was called 'Lotus'. Renault had even been a team in F1 prior to this, between 1977 and 1985, albeit run from a different location. The team in Enstone was originally founded as Toleman, before becoming Benetton.

The Aston Martin Team was founded as 'Jordan', before becoming 'Midland' and then 'Spyker', 'Force India' and 'Racing Point'. All based at the same site in Silverstone. In fact, there are only four teams that have been known by the same name since their founding (three if you exclude Haas who only began racing in 2016), and while Ferrari, McLaren and Williams have always been called such, none of them are now owned by their founders.

Why has it been this way? Why are companies happy to build themselves on top of the ashes of previous operations, rather than starting out on their own? The answer is very simple, because of their *experience*.

Teams built from scratch have a habit of disappearing soon after they appear. In 2010, three new teams were added to the Formula 1 grid (interestingly it should have been four teams, but USF1 was wound up before it had even produced a car). These were all new start-ups. Of these, the HRT team had even been known as 'Campos' when the original application was accepted, before they were bought out before even appearing at their first race. The two other teams went through their own process of changing hands before they all eventually disappeared when the Manor (formally 'Marussia' and 'Virgin') team pulled out in 2016. Don't think that it's only underfunded operations that go this way either. The Toyota team entered F1 with one of the largest budgets on the grid before they pulled out in 2009, without ever registering a race win.

Start from scratch and you are doing just that, but buy a team and you're not only buying their site; you are also buying everything they know about building and developing F1 cars. You don't only inherit their current processes, but everything they have tried that didn't work and all their ideas about how to make the current processes better. This knowledge will be invaluable.

Formula 1 cars are not designed from a clean sheet of paper each year. They will inherit a large number of their parts as direct carry over, while most of the others will take inspiration from the ones on the car from the year before. At each step, there will be small tweaks; to take out a small amount of weight, make them stronger or stiffer. These parts start their development cycles as quite agricultural in their appearance compared to what they might look like on the car today, with years of optimisation behind them.

Unfortunately, the chances of a new team arriving on the scene, having designed their own car from scratch, and performing well straight away are probably very small. The barriers to entry are now such that your best bet as a wealthy investor, is to wait for one of the current teams to begin to struggle financially, before coming in and injecting your own money and renaming with your own brand. Why? Because they will have done a great deal of the learning and optimisation for you.

29

Theory and Application

So far, we have discussed our agent's preferences, our environment and what we are capable of doing in order to maximise our own utility. What more could we want? We are missing one vital element.

What we have arrived at is a static process. Our target is set by the agent, which defines the content of the control process, which in turn creates actions on the environment to give feedback dictating the next steps. After some time of working through the process, the task will be completed and the model will cease to be necessary. This may be a perfectly reasonable way of performing the task in question but if we ever want to be able to get better results from similar tasks in future, we will need a means of learning from what has happened, and a way of applying that learning to improve any future outcome.

Consider the following: let's go back to the dawn of human civilisation. Our job in the community is to move wood that's been chopped down in the forest to our settlement and we have a team of people that we can call upon to help. The control process we will go through is as follows:

1. Once a tree has been cut down and all the branches removed (our measurement), get as many people as we need to lift it
2. Once we have enough people, begin carrying the log towards the settlement, avoiding obstacles wherever necessary

3. Once we have reached the settlement decide on the best place to drop the log and leave it there
4. Return to the forest to collect the next log
5. If anyone is injured along the way, replace them with someone else and return the injured person to the camp to recover

What we have here works brilliantly, the utility the community gains from having all this wood is certainly positive, much more than is lost when people are injured. However, this is not the way this kind of operation runs in the present day; not because the process stopped working, it just became superseded by other, better, processes. Logs can now be lifted with heavy machinery, operated by a single person, before being loaded onto flatbed trucks and transported hundreds of miles. This replaced all of the systems that used manual labour exclusively, with everyone at significant risk of injury.

In the present, our desire for logs and therefore our utility for having wood has not necessarily changed, and nor has the environment (trees still grow in forests) but the options we have for our control process are huge in comparison. What is clear is that we need some way of being able to manipulate and update our control actions as we learn. This is achieved through the process of optimisation.

There is another possibility as well. In our process of learning, we may well reassess our priorities. For example, once we have set up a new hedge fund and made millions for our customers, we might decide that increasing the wealth of the already very wealthy is not as rewarding as the salary suggests it is, and we would prefer to invest our time and efforts into something more fulfilling.

Finally, as we learn, our uncertainty in every area we have discussed so far can reduce, which is good news for us as we should be able to increase our utility more efficiently for the same level of resource. These are some of the ideas that we'll discuss in this section.

Content of This Section

We'll consider two separate topics in this section. These are:

1. Learning - This is where we evaluate what we've gained from our task in terms of utility and knowledge, and how we might be able to apply it to increase our utility in future
2. Optimisation - This is where we decide how we're going to use our knowledge to make changes to our control process or our Utility Function

We'll discuss both steps in turn and how they can combine to make improvements in our processes.

First, we'll discuss the tools we can use to describe the most effective means of learning and the implications this has for our task and the wider world, which will require a basic introduction into the 'Bayesian' interpretation of probability theory. We'll also finish off our look experimentation by covering the ideas behind experimental inference. After this, we can move on to cover the key step of performing inferences on the outcome of our task and how we can document this to be used during similar tasks in the future.

This learning will be essential in our process of optimisation. Here, we'll discuss both discrete and continuous optimisation approaches, which focus on improving the efficiency of our control process and its robustness. We've already introduced the concept of robustness in the process of decision making. In this section we can relate this concept to our methods of optimisation, and we should also spend some time considering how we deal with performing multiple tasks at the same time.

While these ideas will help us to make the best out of what we have, it's only by performing 'exploration' that we'll be able to make giant strides in our capability. To do this we'll have to set aside some of the time we would normally spend exploiting our existing capabilities. Finally, I hope to demonstrate how these methods link to existing, practical methods of process optimisation in industry, and, as always, we'll discuss sources of uncertainty in both our inference and optimisation steps as well as things we can implement that could help to reduce it.

30

Education, Education, Education

How can we describe the process of learning? An expert in Bayesian statistics will tell you that it's all a matter of updating our previously held beliefs.

You may be surprised to hear that there is a disagreement in the mathematics community over what probability *actually* means. We've already talked about the 'frequentist' interpretation relating to measurements and experimentation. In this world view, probability is universal; the probability of rolling any number with a die is 1 in 6, and the probability of getting heads when tossing a coin is 1 in 2. This is probability that we are all familiar with from school. Simplistically, in this interpretation, probability is derived from the expected frequency of each outcome of a particular system relative to the number of tests you have done.

The Bayesians have a different view. They view probability as subjective, i.e. it varies from person to person. Any difference in subjective probabilities between individuals is caused by the differences in 'prior' beliefs and their own experience, which is probably not the way we're comfortable of thinking about probability today, but it can still be intuitive. If we have never seen a coin toss before, how are we to know that the probability of getting heads is 1 in 2? The first coin toss will give us either heads or tails, so we might be inclined to think that doing the same will always yield that answer. It's only by seeing more coin tosses that we begin to see the pattern emerging; that you are probably equally likely to get either one result or the other.

This view is useful in that it incorporates 'learning' into an individual's view of the world. In the frequentist's world view, learning is just a matter of determining the probabilities for each event, and these will be the same for everyone, while the Bayesian view suggests the learning is the building of individual experience which leads them to feel a certain way. Animosity between these groups is fierce but my experience is that both have their uses; each interpretation comes with a different set of tools useful to your work. The frequentist approach is useful for when you can do a large number of experiments in a controlled, cost-effective way, while the Bayesian approach is useful for when you can't do many tests and you have a good idea of the result you expect a test to yield. This is probably a better model for what life-long learning looks like.

Something we need for our Bayesian scheme that we don't need for our frequentist scheme are some 'prior beliefs' for the question we are trying to answer. These prior beliefs might be something we have a lot of experience in, say the likelihood of a mistake being made in a process you deal with as part of your day job, or this might be held in almost complete ignorance of reality. An example of this could be the likelihood of a presidential candidate winning the election when they are the only person to declare that they are running. The form of our prior beliefs can vary significantly and needn't be formally expressed; the learning scheme we're trying to use is one that builds over time to account for all the available evidence. Certainty is not our friend here, as a position of certainty cannot be affected by new evidence that contradicts the position, nor is it likely to be a fair reflection of the totality of the evidence. Some examples of prior beliefs can be as follows:

- I don't know exactly what day my Grandmother's birthday is, but I know it's at the end of June
- Previous evidence suggests that the expected return from an investment is 5% a year with a standard deviation of 3%

- I have no idea whether the republican or democratic candidate will win the next presidential election as history doesn't seem to suggest there is a pattern
- I believe the current parameters I'm using to model a system are appropriate with a maximum error of around 5%

Note that the above can differ between individuals. Someone you barely know will have no idea when your grandmother's birthday is, while your grandfather probably knows precisely. Someone with no experience with the investment you are making won't know the figures above and someone who is looking at results from a different time-period will have derived slightly different ones. Someone with intimate knowledge of public opinion and history might have a very clear idea of who is most likely to win a presidential election. An important point is that there is no expectation that our prior beliefs will be correct, particularly if we haven't seen any evidence yet. They are merely a starting point on which to build when the evidence starts coming in. We expect the evidence will move these around, possibly reinforcing them or possibly contradicting them all together. Your mother might tell you that your grandmother's birthday is actually in January.

Our model of learning for this process will follow the Bayesian interpretation; as we gather data, we make inferences that affect our prior state, and depending on the evidence we see, this can either reinforce our existing beliefs or move them somewhere else. A visual illustration of this is shown in Figure 27, which we can use to help explain some peculiarities of how humans process the information that they get from the outside world.

We are forever reading in the press about political 'echo chambers' online. These are so called because they are visited by people who already hold the views being promoted, and constant exposure to this kind of environment is thought to lead to radicalisation. This is where people see so much 'evidence' to reinforce their prior beliefs that they become basically immovable, as all contrary evidence is rejected by being negligible compared to the mountain of reinforcing evidence that they've seen.

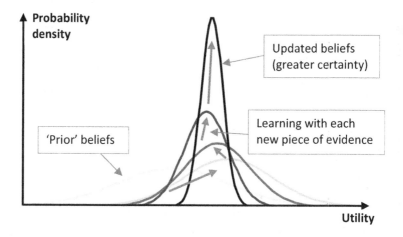

Figure 27. Illustration of Bayesian updating, starting with prior distribution

Indeed, this idea of reinforcement of prior beliefs through experience can help us in explaining cultural differences between different parts of the world. In the UK, we watch news footage of deadly shootings that have happened in the USA, aghast at how a civilised society can hold these attitudes to such dangerous weapons, but we are ignoring the context that exists in the minds of people from each country. In the UK, firearms have never really been owned by your every-day citizen, whereas in the USA, it is commonplace. In some areas, guns form part of the way of life, with gun ranges, gun shows and hunting expeditions being part of the culture. For an avid, responsible gun fan, the idea of a mass shooting that might make the international news seems as alien as it does to the average citizen in the UK, but in their mind it can't be gun legislation that is to blame because of all the experience they have with responsible, law abiding gun owners.

The Bayesian interpretation can also be used to explain a similar phenomenon: 'superstitions'. These can be explained by an individual interpreting the evidence they've seen to spot a pattern that doesn't exist. When you are asked to choose between heads or tails in a coin toss, do you always choose the same way? If so, it is presumably because you believe one result will come up more than the other? Maybe your choice is a function of which way the coin is sitting before it's tossed? In reality, determining which way the coin will land is virtually impossible if there is no intent to cheat. And yet, there is probably something in your experience, maybe a long run of wrong guesses, that sticks in your mind and leads you to think one result is more likely than the other. The Bayesian interpretation accommodates this well. It's not that you are ignorant, as the frequentists would suggest, you've just had experiences that have misled you as to the true probability. In other words, you're wrong, but it's not your fault.

Superstitions have even been shown to exist in other species. In a fascinating experiment, B.F. Skinner fed pigeons grain at random time intervals and, after a while, the pigeons began to believe that their actions dictated when they would be fed. Some pigeons developed strange routines of spinning around in the belief that this would lead the grain to fall into the cages. Maybe they had been doing these actions when the grain had fallen previously (at random) and this had led them to believe they had an influence on the world. Clearly with intervals defined by the experimenter, they had no 'frequentist' type evidence to suggest they had any influence over this, however this behaviour is completely justifiable in the Bayesian interpretation. This experiment has been recreated with humans, and it is amazing how few correctly interpret the rewards as coming at random. There is perhaps a lesson here, we might have "evidence" that our actions have an influence on the universe in a wider sense, but that doesn't mean we do.

If we go to the opposite end of the scale, we find individuals who have beliefs that they give up on too easily when provided with contradictory evidence. In my experience, children have very little skepticism when it comes to things they hear in the playground, and they will come home with some very strange explanations for the way the world works, or what caused a particular event. Adults can be guilty of this too; an idea presented articulately by a respectable figure can seem very convincing. This can be very powerful in swaying our prior beliefs, but do they really carry any more weight than what our child overhears in the playground? In the case of certain tabloid opinion columns, I would say not. I'm sure we know of certain individuals who can be pulled in all kinds of directions with their opinions, which are often backed up with very little evidence, and in these cases, a little skepticism could go a long way. A quote I choose to remember in these scenarios is "We should all be open minded but not so much that your brain falls out". I have found James O'Brien's radio show and books to be very helpful in exposing some of the ideas believed by figures of authority and how these really aren't based on many facts.

Fortunately, humanity went through a historical period of enlightenment in the 18th Century. This process laid the foundations for a critical way of thinking about the subject of 'truth', transforming civilisation from being strictly religious into something that used evidence to create new theories about the universe. It was shown that it was possible to move beliefs away from religious dogma towards theories based on evidence; a process which continues to this day and is the basis for all scientific investigation.

Science is our best friend here. While a lot of people associate science with reams of tedious formulae and tables, in reality, it is simply a process. The best scientific theories are those that are supported by the most evidence; if a new theory comes along that explains reality better, it is adopted, but not before enormous scrutiny.

This constant questioning is what separates science from religious belief, and we should even be able to use the Bayesian scheme to help understand this approach. Both science and religion have their theories of how the universe works and both are open to revision or interpretation. Major religions are frequently updating the stance on ideas like homosexuality, female priests and the frequency of worship. Science has its different interpretations of its theories as well; take the Copenhagen Interpretation Vs Many Worlds interpretations of Quantum Mechanics. The only difference in the Bayesian analogy is the strength of the prior beliefs. In a religious setting, the strength is virtually infinite, with no real room for scientific evidence to move it, while the scientific way of thinking is typically more open to changes. But do not think for a minute that the means science is immune to dogmatic beliefs - when personal gain is involved, individuals will stick valiantly to their beliefs about the universe and discount other theories that contradict theirs.

There is no certainty in these processes; it is very unlikely that the state of science will stay the same in any respect from as it stands today. Advances are happening all the time, so we should be open to the possibility that some, many, or all our currently held beliefs are completely wrong.

Learned Knowledge Vs Experience

One final point on the subject of learning is the difference between learned knowledge (acquired through our research) and experience. In the biological sense, learning is the process of manipulating the pathways in the brain so that we respond differently to a different stimuli compared to how we would before any learning. The brain is built up of billions of neurons that are joined together in a way described as the 'connectome'; signals come in through different sensors and are manipulated through each layer of neurons until a response is derived. We can again refer back to some of our automatic reflexes for examples of how this works in reality. Many artificially intelligent systems have demonstrated that we can build machines that work in very similar ways to the brain; the hypothesis states that it is possible to recreate all the workings of a human brain with only simple components of neurons and connections (which relates back to our previous discussion on Complex Systems, the brain being most complex of all).

In this model, there really is no difference between learned knowledge and experience; if we use science and mathematics to tell us something about reality that is contradictory to experience, we cannot say that the experience is superior to the perhaps surprising results of our research or simulations. Experience is encoded in the brain using a series of mathematical rules and is not magic. It is very useful for spotting what we have overlooked in our work, as it includes all our many mistakes, but it cannot tell us about new ideas, which, by definition, we haven't seen yet. For this we will need experiments. I'm sure we have all met individuals who will discount new ideas in favor or pursuing what they have experience in. The above should form part of our response to these individuals, with the rest of our response covered under the subject of optimisation in the next section.

Now we understand what is meant by learning, we can discuss methods to apply it in a practical sense. Under the following heading, we will discuss what constitutes evidence that we should allow to sway our prior beliefs; a process we will call 'inference'.

Experimental Inference

In the previous section on the control process, we discussed how experiments can be used to understand our reality better. We discussed the requirements of a good experiment for obtaining useful results, and now we have these results, we can start to make inferences. Again, we will consider different types of experiments and how we can interpret the results to 'learn' from them.

We'll start our discussion with simple hypothesis testing. To recap, in this type of experiment, we're looking for the answer to yes or no questions. Does changing a process in a particular way make any difference to the results? Do we have enough evidence to suggest we should change our previous beliefs? Is x better than y?

A useful tool we can use in this process is called the 'p-value'. This is the probability of making a type 1 error (dismissing our previously held beliefs when they were correct) given the evidence recovered during the experiment. You will probably come across this value if doing any investigation into the success of a certain medical procedure. It is common in these experiments to compare the current best treatment to the one being studied and determining the statistical significance of any improvement, by calculating the p-value. This is clearly a useful piece of information as it should help us distinguish between improvements that have come about by intervention over improvements that have come about purely by luck; however, it will never be able to guarantee that we have achieved what we set out to.

Let's use the example of adding a new element to a metal alloy to make it stronger. We have created samples of material both with and without the same quantity of the new element, and we're interested in knowing whether we have made the material stronger or whether there is not enough evidence to show that adding the new element makes the material stronger (note that this isn't the same as saying we haven't made the material stronger). We subject all our samples to tests of strength, with the operator unaware of which samples contained the alloy to prevent biasing. We find that sometimes the samples with the new element fail at lower load than the original and other times they fail at higher load. When we calculate the p-value of the hypothesis 'adding element x makes our material stronger', the probability of seeing these results if the element has had no effect on the strength of the alloy comes out at 0.01. This is a 1% probability, which we can take this as a very strong indication that adding the extra element has made a statistically significant difference. We can now use this to update our previously held beliefs and start manufacturing alloy with this quantity of the new element. Of course, we may have changed other properties of the material as well as the strength, but our experiment cannot tell us anything about this.

The p-value tells us the probability that this learning isn't correct. A commonly used threshold for the rejection of our previously held beliefs is 5%, which seems reasonable except for the fact that if we conduct many experiments, we would expect to see a type 1 error once for every 20 experiments. This may be useful if we don't really understand what we're doing but would like to get something published anyway. We can simply perform an experiment where we don't expect to see statistically significant results many times, and only publish the one where the results go below the threshold of statistical significance, purely by chance. This is called 'p-hacking' and we should bear this in mind if it's taken many experiments to get a positive result.

There are many other important tools that we can use in hypothesis testing to tell us about the quality of our results. For example, we can perform tests on the 'normality' of our results, whether they come from a normal distribution or not, or estimate the 'power' of the experiment. We will leave detailed descriptions of these (and more) to other references. For now, we can start to consider inference from regression experiments.

Something we've probably all been guilty of when plotting lines of best fit in Excel is simply hitting the button to increase the order of the polynomial fit such that the line now passes through all our 10 points before happily publishing this as a suitable model. This is a classic example of 'over-fitting', which is a term that describes the process of fitting our model too closely to the data when we would expect to see some noise on the results. This is bad because it means we haven't accounted for the noise in our experiment appropriately, meaning any predictions we make are likely to be less accurate, rather than more. A simple test we can do is to see if the noise from the 'residuals' (the data we collected minus the value of the line of best fit at that point) is similar to the repeat trials we have performed as part of the experiment.

Let's go back to our material testing example and say we are now looking at the proportion of iron in a steel alloy that will give it maximum strength when subjected to the same test as before. If we try to create the alloy with the same concentrations of each element three times (using the same process), the results from testing these should give us some idea of the noise in the experimental process. We would expect each of these similar samples to be at least infinitesimally different from each other and therefore fail at slightly different loads, be it within 10 Newtons or 1000 Newtons. If when analysing the results of the experiment proper (where the proportion of iron is varying), the line of best fit passes through all the points to within 0.1 Newtons, there is probably some overfitting going on. We should try a different form of equation, which doesn't assume such a complex trend, or go back and collect more results. What we're looking for is that all terms or elements in our model can justify their place being there, that is, removing them would have a negative effect on the accuracy of the model.

The indicators we can use for telling us about the quality of our model fits is a fascinating field of study and is something we should all investigate further if we are keen on pursuing a more scientific approach to our modelling. Again, learning about these methods will not just increase your abilities in the field of statistics and data processing, it can also change how you view the results of your experiments and teach you to be skeptical about your prior beliefs.

The above inference techniques, for both hypothesis testing and regression, are based on a frequentist approach to experimentation. Again, this philosophy concerns experiments where it is easy to perform a large number of tests to gain greater understanding of the noise factors. In reality, things aren't likely to be so simple. It would be great if we could perform controlled experiments to gather every piece of information we might require, but this is likely to be impractical and we'll have to rely on interpreting individual pieces of information to build up an idea of reality over time.

In contrast to a frequentist inference scheme, with Bayesian inference, we start with our prior probability distribution and update it every time we gather some new evidence. We would hope that with enough information, our beliefs will converge towards reality. One of the differences between Bayesian and Frequentist inference techniques is that frequentist indicators like the p-value tell you the likelihood you are to be wrong (the probability your data could have been arrived at by chance), whereas Bayesian indicators give the probability that your beliefs are correct. This is something that gives those following the Bayesian interpretation a great deal of satisfaction.

For an example of how this works in practice, let's take the example of a motor race. Here, the likelihood of each car and driver winning the race is not constant in time. It is going to be incredibly challenging to construct a frequentist scheme to give us the likelihood of our eventual finishing position at any point in the race, because the information that we receive is changing continuously. At the start of the race, the car on pole will probably be most likely to win (they are probably there because they were the fastest) and probabilities will decrease on average as you go down the grid. Not only are the cars getting slower, they will also have to pass the cars ahead of them to win.

As the race starts, information starts flooding in; the cars will be changing positions frequently, particularly if they are on different race strategies (choosing a different number of times to come into the pits), and you will be updating your beliefs on the final outcome. Eventually, as the laps tick down, a small number of cars will become clear candidates for victory, while the others can be more or less discounted. At this point, you can probably infer who the victor is most likely to be; however, it's only when the first car crosses the line on the final lap that all the probabilities collapse into certainty and the prizes are handed out. There have been plenty of final lap car failures over the years and you should definitely avoid opening the champagne until the race is over.

In this example, information is arriving in a similar format, namely the positions of the cars on the track, the gaps between them and how many times they have been in the pits. What if our data is coming from many different sources?

Comparing Apples and Pears

In the real world, we are likely to have to deal with pieces of evidence that not only contradict each other, but come in different forms. Some will come from relevant experiments, others from experience on the job and testimony from other colleagues.

In motorsport, a classic example of this is contradictory information from the driver to what you've measured from sensors on the car. A driver might report a severe loss of grip in one corner that doesn't tie up with anything in the car telemetry. As we discussed earlier, the driver's internal model of the car is extremely refined and the measured data is hardly what you'd use in a robust experiment, with only a handful of instances in any one run. However, the driver is likely to have taken a slightly different line through the corner on every lap and the sensor readings are still more direct measurements of car performance than what the driver is perceiving.

How can we weight these two pieces of information? We need to understand the reliability of each of the measurement techniques. The sensors we have on the car provide incredibly detailed information about specific areas of car performance, but they can only tell us what happened in a limited context and not how it is affecting overall performance. The driver, on the other hand, is part of the loop and should be able to describe exactly what they feel is holding them back, based on their years of experience. Unfortunately, they are not engineers and there will often be situations where they struggle to interpret the behaviour of the car.

How does this help with our learning? Simple; there will be situations where we know the driver always gives reliable feedback – maybe the feeling of too much understeer around the circuit that can be fixed with a couple of turns on the front wing flap? We can give this a very high weighting and allow it to sway our beliefs about the car, meaning we probably won't even consult the sensor data. Alternatively, if we are looking for solutions in a situation where we know the driver tends to struggle to interpret the car behaviour, we can give their feedback a lower weight than what we gather from the data we log on the car. This is because the driver's interpretation has a high 'variance', or a lot of noise. There may be occasions where the driver is particularly adamant; they may repeat the comments over and over again which is like getting a lot of similar data points from a noisy signal. A lot of information that has a low weighting might still be able to sway our prior beliefs, so we may conduct a more thorough investigation. If the comment goes away on the next run, consider it part of the noise.

We can use any of above techniques to update our models for the task, including the magnitudes of uncertainties that are inherent in the processes. This should help us to plan the remainder of the task more effectively, as now we understand more about the environment we are working in.

Inference on the Outcome

The last few examples cover inferences on the results of our experiments and knowledge gained from watching our task play out over time. We'll consider inferences on the outcomes of our tasks under this heading. The final outcome is a point of learning for future tasks in a similar way to the elements described above, but it is also one way of checking our Utility Function. It could be argued that the outcome of the task is the most important inference of the whole process. Here, all the uncertainties that we have been dealing with in our task have reduced to zero and the reality of what you have achieved has become clear.

Let's say we have completed our drive to work in record time and now takes half the time it used to. This has come from fine-tuning all the elements of the process and we decide this is a suitable point to divert our attention to something else. Has this made us as happy as we thought it would? We had put a relatively heavy weighting on the time taken to perform the journey but now we must contend with the extra stress of driving on the motorway compared to a slow but scenic and easy B-road. We thought this would be worth it, but maybe it isn't? Perhaps we find that if we set off at the usual time, when we get to work the office is empty and there isn't anybody who can help us with the tasks we can normally perform. We could instead choose to cash in the extra time at the other end by spending a few more minutes in bed, but if our partner is waking up the time we would have before, it's impossible to get any extra sleep. In this case, we have delivered on what we set out to accomplish but we find we didn't really want to as much as we thought we did. Is this a failure? I doubt it, as we have learnt something about ourselves; our Utility Function has become a little more certain (for the moment at least) and we can apply this learning to tasks we encounter in the future.

We must spend some time trying to understand the process that we went through and what we have received back in terms of resource and utility. This is the part of the task that we have most information to feed back into our optimisation process, so it's important that we make the best use of it. We will thank ourselves when a similar task rolls around in the future!

On some occasions, reflecting on the success of a task will be difficult. This might be because we are lacking a 'control' experiment to compare against. One area in which that this is certainly true is interviewing for a new position in your company. Job interviews are something that we've probably all experienced at some point in our lives. In many cases, they are seen as a necessary evil, with time allocated by both the panel and the interviewee to get a sense of what that person will be like in the role that's available. But does this actually work? After we've offered the job to our preferred candidate, how do we compare their performance against the candidates we could have had?

There are plenty of studies out there that suggest that the job interview approach to hiring is not a particularly good one, partly because there is no opportunity to learn from mistakes. Sure, you may have chosen someone who turned out to be completely inappropriate for the role but that's only because all of the other candidates were worse and you were under pressure to fill the position, right? Maybe this is true but without any feedback, how are you supposed to make improvements to your process? More worryingly, how can you be sure that your own prejudices didn't lead you to choose someone who fit your idea of what the most appropriate person would look like, rather than the best candidate?

Maybe there are things we can do. Occasionally there will be situations where we will need to hire more than one person for similar positions. Now we can look at the performance of the individuals we hire and compare back to our expectations from when we hired them. When we went through the process, there was probably one candidate who we labelled our 'first choice'. Did this individual go on to perform better than the other successful candidates in a way that justified that label in the hiring process? Maybe they did, but if they didn't, you can go back and look for the information you missed when you conducted the interview. Was there one good point that you got too hung up on and this biased the whole approach?

Unfortunately, this part of the process can do nothing to affect the outcome you've experienced but it can help you to avoid the same mistakes in the future. Each of these reflections can help you to refine the shape of your Utility Function, either by reducing its uncertainty, say if you weren't sure whether you were going to enjoy something or not, or changing its shape entirely, if you realise that you just don't like the things that you thought you did. It isn't necessarily that we were incorrect about having enjoyed them in the first place. People's preferences change all the time, and you may just be beginning to realise that yours are as well.

I Should Probably Write This Down...

With our task now having been brought to a close, how are we going to make sure we retain as much of our learning as possible?

We're going to have to produce some 'documentation'. We should be clear that when we say this, it needn't mean reams of paper detailing every step that has taken place in the task. In fact, this is not likely to in our best interests at all, given that sifting through all the information that does not include any learning is inefficient for future tasks. Instead, we should emphasise important pieces of information, things that were not obvious at the start, but were learned during the process. Routine elements of the task, such as the control process used, should take up minimal space in our documentation as they are not necessarily adding anything; we need only include so much information as to be clear what it was we did.

In terms of how this should be published and stored, we need to consider how it might be used in future - if the task is one that takes place infrequently, with ample opportunity to forget any learning that was done, some kind of physical documentation wouldn't go amiss. This could take any form, from a post-it note on the tool that you only use for this task, to a voice recording that you keep on your computer or a formal report with appendices. If the task we were working on only involved ourselves and we perform it frequently, we are likely to remember any surprises we had, meaning physical documentation may simply be inefficient.

Efficiency is the key metric we are looking for when judging the effectiveness of our documentation; we are not improving the process directly by recording what we have done, so time spent documenting is dead time in that respect. Something that is efficient to compile and efficient to read is therefore optimal for the task we are conducting. The medium that can get the most learning into as little time to digest at the other end will be the best form of documentation for us.

We're covering this topic in this section, but strictly speaking it should live in the control process element of the model - if it were to live outside the control process, it cannot be part of our optimisation step (that we'll discuss next). Optimising the types of documentation and content we produce is key for efficiency. As we have said, we should spend as little time on this as possible to communicate the necessary information.

Is it possible that no one will read the reports we produce? Yes. Is it possible that we never look at the reports we have produced again? Yes. Is it possible that the reports get forgotten about all together? Yes again. However it sounds, these are not good enough arguments for ignoring it altogether. We should consider the alternative; producing no documentation at all means all the information lives only in the heads of those that performed the task and anyone they told about it, relying on every detail being remembered. This feat of memory is unlikely, and people may even misremember pieces of information and end up with the wrong idea entirely. By producing the documentation and storing it in an appropriate library means everything that was learned is available for all to see, which only help in improving our effectiveness. We can take all the lessons from previous work and build on them the next time we do something similar, rather than being stuck in a cycle of continuously relearning what we have forgotten.

Of course, this relies on us seeking out and reading the documentation we produce at the start of tasks we think are similar. This should be the first step in our optimisation process, and if we find this step is not yielding improvements, we should probably go back and consider the form and content of the documentation we are producing.

Uncertainty

When we discuss uncertainty in our inferences, we are trying to understand the probability that something we have learnt is not correct. This is not a good position to be in. We could do with avoiding spending a great deal of time and effort on models and processes that do not represent improvements.

When it comes to experimentation, we might be in a position in which we've calculated the uncertainty in our inference precisely; for example, in experiments where we calculate a p-value. This tells us precisely the probability that the results you have gathered have occurred by chance. While a low p-value is strong evidence that what you have changed has had an effect, it is no guarantee (the Bayesian schemes have their own equivalents for these values).

In other cases, where there is a high probability of seeing the results we got just by chance, we cannot assume that whatever we are testing has made literally no difference. Here, we're risking a type 2 error, which is that whatever we're testing *does* make a difference, but our technique is too imprecise to pick it out. To understand how susceptible our results are to this, we need to establish the power of the experiment; a test with low power is always unlikely to give a positive result, so you shouldn't necessarily take the results you get as being true in all cases. This is another measure of our uncertainty in the results and is something we would benefit from knowing.

Most of the time we will not have the benefit of a rigorous experiment on which to base our learning. In these cases, our best chance of understanding the uncertainty is through the method of experimentation or information transfer. If we hear that our company is considering redundancies from one person who we know to be unreliable, we should treat it with caution. If we hear the same thing from multiple individuals whom we would normally trust, we can probably be more confident.

With this approach, we can learn from information that comes from multiple sources. If we have some sparse experimental data, some empirical evidence from literature and testimony from trusted colleagues that all point the same way, the chances of it being incorrect should seem small. If each of these sources point in different directions, we can weight them based on the individual likelihood that each is correct.

The problem of reducing the uncertainty in our learning is the whole purpose of science. We can look to mathematics, and statistics in particular, to give us tools to describe it, however we will never be able to remove it completely. At the very best, we can hope to reduce the chances of being wrong to a level so low that we won't benefit from considering the alternative view. We will not get there easily though. The scientific world is filled with contradictory explanations for different phenomena, each with a huge mass of evidence supporting their position. There will only be one theory that explains the workings of universe, and it's unlikely to be anything that's being considered today.

31

Optimists, Not Perfectionists

If you have ambitions in your career, turning up and 'cranking the handle' all day before leaving probably isn't going to be enough to get you noticed. You will need to look at the jobs you are tasked with from the outside, see where the problems lie, and then use this information to create a series of actions that can be implemented to make the process better. We will call this process 'optimisation'.

This is one of my favourite parts of my day job, as for me, it separates those who truly understand *why* they are doing their job from those who are merely good at executing it. If you can take a process and make subtle changes to arrive at another that regularly produces better results, you will have added value that far exceeds the improved outcome of a single task; for example, improving the method of using the simulator to tell you how to develop your car will have payoffs that far exceed any single development that passes through it.

To achieve this, we must be critical of everything we come across in our processes, seeing optimisation as a state of mind that has us continually questioning our reasons for doing what we do, and the way we do it. If I had a slightly different tool, could I do this faster? Is this step necessary or is it wasted effort? How can I do this better?

Our ability to optimise is what separates us from other animals. Modern human civilisation is built on a process of fine tuning how we approach every aspect of our existence. Why build a fire when you can use an electric oven to cook your food? Why walk long distances when we can drive? Why accept a shortened life expectancy when advances in medicine could help you live to one hundred? We know that this way of thinking will exist for as long as there are humans, and the world one hundred years from now is going to look very different to the world today. Before humans arrived on this planet, there weren't many occasions where you could say that with as much certainty.

In many ways, the whole of this book has been about optimisation. After all, why would you be reading this if you weren't interested in making improvements to the tasks you participate in every day? Having a greater understanding of each of the elements of our model will only help us if we can use them here, in our optimisation process.

In terms of our model, we'll describe 'optimisation' as a method of improving our control process such that the expected utility we get from the outcome is increased. Our true goal is to get to a point where we can do no better with the resource available to us. We will call this solution 'optimal'.

We must be clear that by 'optimal' we do not mean perfect. Perfection in the context of control is somewhat meaningless. Surely a state of perfection would involve completing the task with no resource spend at all, be it measured in money, time or anything else? Optimal, on the other hand, is something we should be able to get close to. For this we will need to have a clear understanding of all the processes involved in achieving our outcome, and the utility that can be obtained for their successful execution.

Reaching our target is unlikely to be achieved overnight. We should expect our optimisations to continue for as long as we are performing the task we are interested in, and the likelihood is that there will always be a better way of doing it. We can use the analogy of Evolution by Natural Selection at this point; life on Earth has been evolving for billions of years, with each step making a species slightly better adapted to its environment and therefore having a greater chance of survival. This is how we should view optimisation of our processes, although hopefully we can expect things to progress slightly faster.

We can look at examples like the manufacturing industry to demonstrate the progress humanity has made for what is fundamentally the same task. At the dawn of civilisation, tools were made by hand, whereas now we have huge factories in which large machines create products with little human intervention. The scale of the process dwarfs what used to be possible in the past, and we would be naïve to think that these improvements won't continue long into the future. It will be through this process that those changes occur.

If we go back to our log moving example from earlier, we can start to explore this idea further. Imagine we've had a bad year. We've suffered a great number of injuries and logs have been piling up in the forest because there haven't been enough people around to move them. This has been identified as a severe weakness and everyone is pretty dissatisfied with how it's all is working. We've convinced the village to invest in a couple of experiments: first, we tried clearing one of the routes between the forest and the village of all foliage that was preventing progress. We then tried building a trolley out of some of the wood, in the hope that this would speed up the process, making it possible to carry more logs on each journey for the same number of people working.

Both of these experiments turned out to be overwhelmingly successful, with the resource lost keeping the paths clear and the trolley in working order more than made up for with increased productivity and a reduction in injuries. This will increase the utility gained from the task significantly in the village. We decide to carry both of these improvements forwards into the control process we use for each task. Here we've applied our learning to make our processes better; by moving resource around, conducting experiments and feeding results back into the process, we've made a real difference.

Like our steps in the control process, we can split optimisation into discrete and continuous methods. Discrete optimization will concern quantities that only take integer (or 'quantized') values, which involves moving resource around in blocks. We cannot naturally talk about hiring 3.4 people or buying half a computer for example.

Continuous optimisation refers to tuning of continuous values of a process to get us into a better position than where we started. If we go back to our example of climate control in a car, do we get better results when we increase our resource (fuel) allocation to this task in order to keep better control? Or are we better off relaxing the control slightly and having less fuel spend? These improvements will be judged by whether or not they increase the agent's utility.

Finally, we can consider any process of innovation as a form of optimisation, where improvements will come about from a paradigm shift in the way we think about the task. Going from no trolleys carrying logs to having one is a significant shift in the way we approach the task; a method of optimising which will require a great deal of creativity.

This process of optimisation is likely to require resource from the agent to perform. This process will certainly take time away from the normal control steps, but it is likely to take financial resource away as well, which is something we must bear in mind. There will come a point where we are not able to perform our normal process because we've dedicated too much effort to our optimisation - getting the balance right will be key to our final outcome.

What Should We Optimise?

Cast your mind back to when we first introduced the concept of the Agent. We described them as being made up of two parts: the Utility Function and their resource. The Utility Function is there to tell the agent what they want, and the resource is what they have with which to achieve it. When it comes to making decisions about how to act, the agent will probably have to sacrifice some of their resource in the hope maximising their utility. This resource loss is going to be painful, so we hope that the utility gained from whatever it is they're doing will go some way to compensating this (you clearly wouldn't spend time and money on something that you expected to lead to a bad outcome). What we have here is a way of grading our task. By comparing the resource spend to the expected return, we arrive at the task's 'efficiency'.

Efficiency will be the prime target for our optimisations; we want a process that gives us the biggest return in terms of utility, for the minimum resource outlay, in other words, as 'lean' as it can be. When we're looking to buy something we need, we want to spend as little of our financial resource as possible. If what we're doing requires us to spend time doing something unpleasant, we want this be over as quickly as it can be. When we design experiments, these need to give us the best chance of achieving good results for the resource we invest in them. Similarly, we need to weight our resource spend on the control process and optimisation according to how likely we are to get improvements from our optimisations.

Once we've set up our control processes we should be able to see how each step influences the inputs to our Utility Function. As we've discussed in our product design examples, staff satisfaction will be hurt if we're trying to cram too much into too short a timeframe, customer satisfaction will fall if we have low expectations of the staff and revenues will suffer when delaying processes so that competitors can get the upper hand. What we're seeing is that the maximum expected utility from a particular process is going to be a balancing act between many competing inputs. This is to be expected.

In order to get the most out of the processes we've designed, we will need to understand how sensitive they are to the allocation of resource. If we take money and staff away from the HR department, you can expect staff morale to reduce. If you take money away from the design department, you can expect the quality of your product to suffer. Somewhere in all these trades will be at least one optimum point where you would expect to be getting the most out of everything, which will mean making sacrifices in some areas that are performing well to improve areas that are performing badly. We cannot expect perfection, but we can expect the *optimum*.

Again, this is an example of our holistic approach in action. A reductionist optimisation might focus on a particular area that's struggling to maintain standards and reallocate resource from an area that is performing well. If the area we take resource away from is critical to our operation, while the struggling one is of minor importance, we're not going to increase our utility overall. Our holistic optimal does not make any claims about the performance of individual areas; it merely finds the best distribution of resource to give the best outcome. If one area appears under-resourced, it does not mean moving spending from other areas will make things better when considering the task in its entirety.

Robustness

Aiming for maximum efficiency is a sensible starting point but, as we've discussed several times, we are subject to all kinds of uncertainties when we look at these problems. There are those that arise because our targets are uncertain, those related to our modelling inaccuracies and so on. When we design a process, do we really want to consider only the expected outcome? What if we had designed a process where we believe there is an 80% chance of making a healthy profit but a 10% chance of bankrupting our entire company? Is that acceptable? Clearly, we should consider the impact of what we design if things don't quite go to plan. We need to add some 'robustness'.

The future is always going to be uncertain and we can refer to our trusty driving example to understand how this might affect our decision making. Let's say there are two possible routes available to us; we've travelled both enough times to know that route one is likely to be 2-3 minutes faster on average than route two, however, there is a level crossing on that route, which if closed at the wrong time will add 10 minutes to our journey. We estimate that the level crossing is down on 20% of the journeys we take. Route two is usually slower but more reliable. Let's say there is only a 1% chance that this route will take as long as route one when the level crossing is down.

Now let's say you have a very important meeting to attend first thing in the morning - which route would we take if we have enough time to take either, but we know we will be late if the level crossing is down on route one (or equivalent for route two)? Yes, route one usually gets us there slightly faster, but surely the utility lost from being late to the important meeting will dominate any small gains from being 2-3 minutes earlier? In this case it seems sensible to take the slightly slower but more robust route two.

We can explain this with the help of the diagram in Figure 28. Here, we have two probability distributions for expected outcomes. 'Option 1' (solid lines) is the option with the highest expected utility, but if we want something robust, i.e. the likelihood of getting a bad outcome is low, 'Option 2' (dash-dot lines) is a safer bet. Here, we've changed the way we would normally behave to reduce the possibility of disaster.

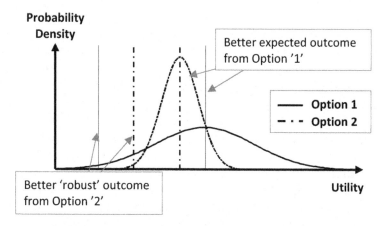

Figure 28. Illustration between 'maximum expected' and 'robust' utility values

We see choices like this in industry all the time. We can take on a lot of debt in order to build the business, release new products, hire new staff or whatever it takes in order to grow. If all goes well, we'll probably be fine. The revenue from the new products will roll in and our staff will go home happy that they've been paid. But what happens if that first product doesn't go to plan? Maybe a competitor surprised us by moving in early to beat us to market? Now we can't pay off that debt as quickly as we'd like, spending starts to drop, and the staff start to get very uncomfortable, which could very well bring the whole thing crashing down. We clearly all have our own tolerance for risk but there will surely be a point where everyone is willing to sacrifice a small amount of possible gain to reduce our chances of total failure?

When we come to our process optimisations one thing we can consider is maximising the minimum utility we expect to gain from the process, which is similar to our 'maximin' approach that we presented in our discussion of decisions. For processes with a continuous spread of possible utilities, the concept of 'minimum' possible utility doesn't have much meaning. Let's not forget that there is an infinitesimally small chance of being hit by a meteorite on your way to work (this would probably spoil your day). In this case we can choose a lower limit, say the point of 10% probability on the lower end of the outcome range, and increasing this point instead of our expected utility might not have you maximising your process to the fullest, but it should help you to avoid incredibly bad days as well.

Multiple Tasks

Of course, the task we're optimising is not likely to exist in isolation from all others; within our optimisation loop we must consider the impact our task is having on the others that are competing for our time and resource. When we discussed material and time resource, we considered the opportunity cost of the task we are performing, and again, these are resources that can only be used once. If we choose to complete a task that is reasonably inefficient for utility gain, we will miss out on opportunities where we could have done better.

We need to account for this in our optimisations, which could involve performing optimisation on several tasks concurrently. This will be more complicated than only considering a single task but may be essential to prevent resource being spent inefficiently. When we create project plans, such as Gantt charts, we are usually mindful of other tasks that are being conducted at the same time.

Imagine we are designing two products, but we only have the staff to perform one effectively at any given time. We could choose to put a proportion of the staff onto each project, but this will detract from the progress of the other. If one is thought to be worth significantly more utility than the other (maybe one project is a revolutionary new mass market idea and the other was for a niche sector that was already saturated), we would be wise to dedicate most staff to the more valuable task and move them across to the other when this task is finished. If both tasks are thought to be worth more or less the same amount, we could decide to do one of two things; we could split the resource evenly, in which case both would be finished at about the same time but take longer, or we could perform one after the other. In this instance, performing them one after the other is likely to bring some of the utility gain earlier in the process when the first task is completed, which is probably the rational decision if we have some weighting towards instant gratification. For greater numbers of tasks, the logic still holds but the optimisation increases in complexity again.

We can describe our skills resource as being limited in a similar way. The skills within our agent are likely to reside within certain individuals or groups; we could split resource for two projects in half, so we perform them concurrently, however, if we only have one person who has a particular skill that is needed for both projects, we're going to have problems. We are likely to find a bottleneck around that skill, where things take twice as long as they could do. Avoiding this could get expensive...

Considering the multiplicity of tasks we could be involved in leads us to the natural conclusion of our holistic way of thinking. If we want to maximise our utility we need to consider not only the task we're studying but also anything we're doing at the same time, plus everything else we could be doing instead. This level of rigor in our optimisation may end up being immensely impractical, and we can probably do much better than we're doing now by only considering a few options. We shouldn't lose sight of this objective, though. We're never likely to reach a state where we don't want better.

In this chapter we've covered the targets for our optimisations. We want our task to be as efficient as possible with the minimum resource outlay for the maximum utility in return and, at the same time, we want to minimise the probability of getting a truly terrible outcome by adding robustness. This must all be performed while being mindful of the other ways of we could be spending our resource, or the other tasks we are performing in parallel.

This is the goal of an F1 team; the available resource constrains what's possible, but we can at least hope to extract the maximum out of it. How we apportion material and human resource between tasks like wind tunnel testing, vehicle design and race support will have a strong bearing on where we finish during the season. Any time and money wasted is resource that another team could have spent making their car faster, and we need to make sure we minimise wastage like this.

Achieving maximum efficiency could even help us overcome teams with more resource, and there is already precedence for this level of performance. As was mentioned in our introduction for this section, the Toyota team was once known to be investing more than any other; they left the sport a few years later without ever having registered a race win.

While efficiency is clearly desirable, we must avoid increasing the risk of total disaster. Taking out a huge bank loan that we're only able to pay back if we win the championship may increase our expected finishing position, but it could lead to the end of the team if we're not quite as good at spending it as we thought we were. The shareholders know that there is little hope of their investment paying off if they continue to shovel money into an inefficient enterprise, and they are naturally reluctant to commit money that they don't expect to get back.

With these ideas, we have a reasonable guide for what we will need to optimise. Now we can start to consider how we're actually going to achieve it.

32

Race to the Top (Or Bottom)

Optimisation is one of the most widely researched areas of applied mathematics, and it's not hard to see why. Has there ever been a task in human history that wouldn't benefit from optimisation? It's likely that there never will be. Developing the tools that can help you to achieve it is immensely valuable.

As with most of the subjects we've discussed so far, this topic is broad and we'll only be able to scrape the surface here, focussing on a handful of techniques that mathematicians have come up with to help address some of the most common problems. While we're discussing these techniques under the heading of control process optimisation, there's nothing to stop us putting them to other uses as well. If we were designing a new part for an engineering project, we may well have an 'optimise design' step that draws on some of these techniques.

In each of the examples we're optimising parameters in our Control Process; these will normally be quantitative but could be measured on continuous or discrete scales. Some examples could be:

- How much should we charge for our new product?
- How many people should we dedicate to each area of the task?

- Are we better off spending more time on the design of the product or the manufacturing process?

Think of answering these questions as like trying to map an area that you've never been to before. In order to become a famous explorer, you're going to have to find the most important points on this landscape, that is either the top of the highest peak or the bottom of the deepest trench.

The ranges of each parameter that we could conceivably use describe the 'search space' that we'll be looking through; this is like the boundaries of the map we're optimising over. On top of this search space will be a function that describes the quantity we are interested in optimising, be it expected utility, utility at a robust level or otherwise. This is like the contours of the map that describe the landscape. Depending on where we're looking, this could be a gently undulating plain, or a rocky mountain range, full of jagged features and cliffs.

Our goal will be to find the single point on this function that maximises the outcome from our perspective. The coordinates of this point will tell us our prediction of the parameters that lead to this optimal outcome, and this is what we should take forward and apply in our task.

While we'll be describing mathematical approaches to optimisation, you may never need to use any of these techniques in a formal way. What I hope they can do is stimulate some ideas about how your optimisations can be achieved practically. I encourage the interested reader to seek out other references if these descriptions are too light on detail.

For now, let's start simple...

Simple Optimisation

For most informal tasks we participate in, we won't need to consider any complex algorithms to optimise how we are going to behave. In these instances, we will simply have a few choices we can make that relate to how we are going to perform the task.

This kind of optimisation is going to be relatively simple; we just need to lay out all our alternatives for how we could perform the task and choose the one we expect to maximise our utility. This isn't something that will take a lot of explaining but it is definitely a valid method for arriving at an appropriate solution.

While this seems simple in principle, we must remember to be holistic in our thinking, as the expected utility for a project is made up of all factors that form the Utility Function. Performing a boring task estimated to take many hours in one go is unlikely to maximise utility if we're considering the health and well-being of those performing it.

We can consider a simple example of planning a visit to see different sets of friends at the weekend. We'll imagine that they live in different cities, so we'll need to do some travelling between them, and we can choose how much time to spend with each and on any combination of Friday evening, Saturday or Sunday. It would also be nice to have some time to yourself at the weekend to recover for the following week.

Friday evening might be shorter, and you may be tired from a busy week at work, but it's also the time when most things are open and the atmosphere should be better than on Sunday, for example. Saturday is a good option but if you've had a long night on Friday, choosing the following morning to meet the next group of friends might not be very enjoyable. This condition will need to form part of the optimisation, as going out on Friday evening with one set of friends will change the expected utility of a visit to the other on Saturday, particularly if this would require driving early on Saturday morning! Visiting on Saturday and Sunday would be a reasonable option but won't leave any time to rest in preparation for the following week.

We are beginning to see that there are some conflicts that we're going to have to resolve. When you've done this kind of thing in the past, you've probably gone through all the possible permutations and eliminated the ridiculous ones straight away, leaving you with a choice between one or two that come close. You may have even tried to 'score' the different options, and we could imagine using something like the 'even swaps' table to come to a decision. Or maybe you decided that the plan was unworkable and rescheduled meeting one of the sets of friends to another weekend.

There is another option as well. If you decided to see one group of friends on Friday evening and it was much more fun than you thought it was going to be, you might be feeling reasonably awful first thing on Saturday morning. If you have agreed to meet the second group on Saturday afternoon and it's going to take a while to drive, this is bad news. You could take this opportunity to review the current state of the task and do another round of optimisation, now that you have more information (the 'measurement' of how you are feeling). You might decide it's now best to postpone this meeting to the following day, which might not give you much time to recover for the week ahead, but it sounds like the utility of forcing yourself to the next event is not going to do you much good either.

Whatever the outcome, it is the process we have been through that is most important. By working through the alternatives methodically, considering all the benefits and drawbacks, we have a come to a holistic decision that should get us close to the best possible outcome - we can even choose to review the decision during the task and make changes if necessary, as new information comes in.

We can use this way of thinking when tackling more complex optimisation tasks. We will consider these under the following headings.

Continuous Optimisation

If we are so motivated, we will be able to come up with something a little more sophisticated for the task we are embarking on. Think of the techniques under these headings as the satellite navigation to your problems of finding the optimum in the given landscape. Here we will discuss several methods that will help us when resource allocation varies continuously.

- Analytical optimisation

In a few special cases, we may be able to derive the optimum point analytically. This will be possible if we have a small number of simple functions of how utility gained varies with resource allocation.

Figure 29 shows three lines of utility plotted against the selling price of a new product. The curve labelled '1' shows the utility gained from the expected profit when selling at each price. The curve labelled '2' shows the utility gained from customer satisfaction at each price respectively. As we can see, the customer satisfaction utility increases as we go towards zero (giving the product away for free) but the utility gained from revenue reduces significantly as we get to zero, going negative when we would be selling these products at a loss. There is a maximum in this curve, which is the price that gains the highest profit for the company. This is the point where the increased margin on the product begins to lose out to a reducing number of sales from lower demand.

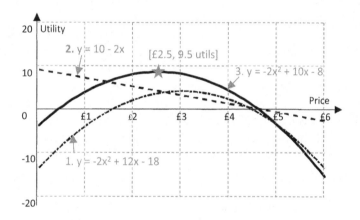

Figure 29. Analytical optimisation example

By adding the two curves together we get a third curve, labelled '3', which has an optimum value that is below the optimum for maximum revenue, but is significantly above the optimum for maximum customer satisfaction (£0). We will need to do a bit of maths to find the precise optimum (please refer back to your differentiation notes from school!) but this will give us the answer without having to resort to more complex algorithms. In this case the price we should be setting the product at is £2.50 which is the price of maximum utility for the business - high enough to make a good profit but low enough to ensure the customers are very satisfied with their purchase.

- Gradient-based optimisation

As we start to increase the number of variables we're dealing with, solving for the analytical optimum becomes less and less efficient, particularly if we don't have nice mathematical relationships between inputs and outputs. This is where a gradient-based optimisation can help. These procedures start by choosing one or more starting points and evaluating the gradients of our function across all dimensions, and once the algorithm knows which way the inputs must go to increase the overall utility for our process, it takes a reasonable step in that direction and starts to evaluate gradients again. With the algorithm now in its new position, it can either decide to continue in the direction it was already heading in or change course to reflect a change in the landscape at the new point. If by taking a step in the optimum direction it has found utility has reduced overall, it can choose to go back and try a smaller step.

These functions can 'home' in on the optimum over the landscape like a marble working its way to the bottom of a bumpy trench. There are risks, however; as is the case with the marble, the optimisation can get stuck in a local maximum/minimum, where it is unable to escape to get to the true optimum. An example of a function exhibiting many local minima is shown in Figure 30. If the optimiser finds itself in one of these local minima, it will assume it's reached the global optimum, as it has no concept of the space outside of this.

This problem can be remedied by launching the optimisation with more starting points. You would expect that if these cover enough space, at least one should avoid all the local optima and reach the true optimum for the functions you are studying.

It's relatively straight forward to implement 'constraints' on this kind of optimisation. For example, if you're deciding how to spend money across all of your departments, it's no good for the optimiser to tell you that the highest utility can be gained by spending the maximum possible budget in all departments if the total then exceeds the amount you actually have to spend. More rationally, we can set up a constraint that says the total spend across all departments cannot exceed the budget you have, which should return a much more sensible recommendation.

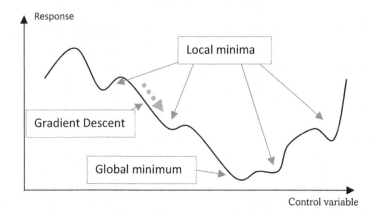

Figure 30. Examples of Local and Global Minima

One limitation of the technique is that we need smooth functions for this to work; if our functions are noisy, with many local maxima/minima, we will be forever getting stuck with gradients that are pointing us in very strange directions. For noisy functions like these, we can consider other types of search methods.

The idea of gradient-based optimisation highlights a key concept called the direction of 'steepest ascent/descent', which describes how we should weight all the parameters we are optimising, relative to their ability to improve our outcome at the position we find ourselves in. We may be in a position where the value of a single parameter is dominant in our decision making, meaning making the correct decisions on where to put that parameter is much more important than if we made some oversights when setting the others. Take cooking a microwave meal, where the time spent in the microwave is probably the only thing we could get badly wrong in the whole process. If all our parameters have an equal weighting, we must be more careful in balancing how we deal with each, and we can use this to guide our decision making. A better example of this kind of problem could be in cooking a Michelin-star meal in a restaurant, where there are possibly hundreds of processes to get exactly right or the whole thing is ruined.

A heuristic that we can use to help us is that changes at the largest scales are likely to have the biggest effect when making decisions about the task outcome. This relates back to our hierarchy of Complex Systems that we discussed in our section on the environment. If we want to change the behaviour of the economy, our best chance is to manipulate the largest scales, i.e. the individual markets and the companies. Changing the behaviour of individual employees is unlikely to have as big an effect.

Similarly, if we want to extract weight from our racing car, we are probably best starting off with the heaviest components; the chassis, the engine, the gearbox. Starting with very small items might lead to big improvements in their percentage weight but this is not likely to build to very much in the context of the whole car.

Something that might break this rule is if we have scenarios that are very chaotic. In these cases, changes at small scales can have a large effect on the outcome. Imagine our economy example again but instead of manipulating taxes and interest rates at the largest scales, we give a small loan to a very bright individual as investment in their idea. This might turn into a product that dominates one market and completely transforms the economy. This is, of course, unlikely. The problem with chaotic systems is that they are incredibly difficult to predict, due to their sensitivity at these small scales, but events like this are possible.

- Search methods

Search methods, like 'section' searches, 'line' searches or 'simplex' searches are more robust when it comes to noisy functions, as they are not reliant on gradients to work properly. They work by taking big steps around the space they are searching, gradually getting smaller until they home in on the global optimum.

To visualise how this works, consider the golden section search in Figure 31. In this case, we are searching for the global minimum of the function. Note that the algorithm doesn't know what the whole function looks like; if we knew this there would be no need perform a minimisation.

We start by evaluating the functions at the boundaries, which are labelled as point '1' and point '2'. Once we have the values of the function at these points, we create another point between the two (in this case we've divided the line according to the 'golden ratio'), and use the three values and their positions in the space to decide where to search next, using some simple logic. As we repeat this process, we move around the space until we reach point number 6, which is very close to the global optimum. We will probably need to do some evaluations close to this point to check we are at the minimum, but hopefully the whole process hasn't taken very long. Note how the shape of the function is very noisy; our gradient-based methods would find getting to the global minimum very difficult indeed.

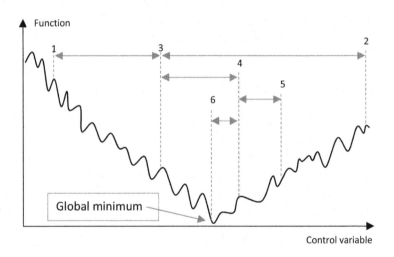

Figure 31. Illustration of golden section search – minimiser moves closer to the global optimum from point 1 to point 6

Line searches and Simplex searches work in a similar way over multiple dimensions. In the case of simplex searches, you can search through as many dimensions as you like.

- Monte-Carlo Simulation

If all these methods have let us down, there is a more heavy-handed approach we can try called the 'Monte-Carlo' simulation, which is a very grand name for a relatively straight forward idea. All we're really doing is generating a large number of 'random' test points, evaluating them for expected utility and seeing which ones come back looking best. If we find that the original set of test points is too coarse, we can refine the search area down and set off more tests. It is very unlikely that we will reach the true global optimum with this method, but we might be able to get quite near to it. We can then use some kind of model regression to understand what the space near the global optimum looks like to find some more points to evaluate.

This method is of course reliant on evaluating the functions over a large number of points, so if this process is complicated (using many simulations for example), we might need to consider alternatives, however it is a useful 'brute-force' method for getting to your answer without risking local minima and other issues.

Discrete Optimisation

While the methods discussed in the above section can be used on continuous functions, we will run into problems if we're dealing with discrete items; we're not going to be too happy if the optimum amount of people we need in our team comes out as 4.56.

Discrete optimisation is slightly more complicated because we are restricted to looking at 'quantized' values when searching through the space. The term 'gradient' starts to lose meaning here as we can no longer really treat the points as being joined together, but the following might be able to help us out.

Our Monte Carlo approach is still valid, as we can restrict our test points to ones that are possible from our ranges of discrete values. This is good news, but this is not necessarily the most efficient way of finding our optimum.

Another approach is to make our function look continuous to start with and find the region that our global optimum is in. As pointed out, this might initially come back saying we want half a person or fork-lift truck but once we have this point in the continuous space, it's reasonably trivial to look around at acceptable points to see which will maximise our utility.

Alternatively, if we only have a small number of discrete options, say we're choosing between three or four people in our project team and whether we need to buy a new milling machine or not, we can set off continuous optimisations for each of the discrete options, as in this case there are only four (3/no, 3/yes, 4/no and 4/yes).

Hopefully with the above options, we can find something suitable for our optimisation task. At the very least, we're comparing different alternatives and selecting the one that looks best, which has to be better than continuing as if we won't be able to do any better.

Innovation - Exploration Vs Exploitation

A final question worth exploring is how do we approach the unknown? There were times in history where parts of the planet were completely uncharted, but it didn't stop explorers venturing into them in the hope of making new discoveries that could bring them fame and fortune.

If we think of the world in 100 years' time, it's definitely not going to look much like the world we're in now. We know this because this has held true for centuries and there are suggestions that the pace of change at this point in history is faster than it's ever been before. How is this change going to come about? Certainly not through doing things the same way over and over again, making small changes here and there – we're going to have to do some 'exploration'; a term we'll use to cover investigations that we are unsure of the outcome of. For example, if we do some experiments on the proportions of every element in a particular metal alloy, we would hope to find how the properties change as we vary the quantities. But could we do even better with a completely new alloy?

These are the innovations that are likely to be of biggest benefits to us, and again, we can look at historical examples. Christopher Columbus set sail to find a new route to Asia while his counterparts were content with trying to optimise their existing routes. This voyage may not have ended as planned but he happened to stumble upon what would become two new continents, with new land and material resources to exploit. Henry Ford took one look at the production methods for cars and thought he could do much better, inventing the concept of automotive mass production in the process. This put him years ahead of his competitors in terms of efficiency. James Watt invented an entirely mechanical means of travelling around the world powered by steam, something quite unlike anything that had come before. These are major successes of an exploratory way of thinking that have changed the world.

While we may not be able to have quite the same impact, we should expect to have success on a smaller scale by incorporating some exploration into our plans. When considering our journey to work, maybe we're able to do the same journey car sharing, or by train? This might bring benefits that far outweigh the benefits of shaving a few minutes off our journey time by car, as now perhaps we can do our e-mails on the journey in, or there might be environmental benefits.

In the world of motorsport, some of the biggest gains have come from approaching the design of the car completely differently. There are plenty of examples of innovative thinking throughout the history of the sport; the Brabham 'Fan Car' generated downforce by sucking air out from underneath the car using a large fan, powered by the engine, and the Tyrrell P34 was distinctive in that it had six wheels, rather than four. The Williams FW14B had active suspension that could position the body of the car in a way that maximised aerodynamic downforce in all parts of the lap. All these innovations gave the teams a tremendous advantage at the time, so great that the governing body promptly banned them all not long after they first appeared.

Solutions like this require some creative thinking, rather than just optimisation of the same old methods. The processes involved in creativity have been studied thoroughly and there are certainly things we can do to promote it within the task we are considering; we can create a positive attitude by doing things like giving serious consideration to new suggestions, setting 'hypothetical scenarios' and conducting exercises in expansive thinking.

When it comes to deciding whether to explore new solutions or stick with what we know, there are certain types of mathematical problems might be able help us. These are referred to in the literature as 'bandit' problems.

Imagine that we walk into a new casino that is filled with slot machines. We know that the probability of winning the jackpot is going to vary between machines, so how do we know which one is best to play? We will have to sit at each for some time to understand the probabilities, and when we've found one that pays off reasonably often, do we stick at this (exploit) or carry on looking (explore)?

An approach that's been designed to help answer questions like this is the so-called 'Gittins index'. This is an index given to each possible option based on the rate of success and failure of each option, whether we have tried it already or not. The choice with the highest Gittins index is the most logical option to try next; the major assumption of this solution is that future payoffs are discounted at a constant rate relative to the current choice. For a discounting factor of 95%, the reward at the next decision point will be worth 95% of the current one, while the payoff at the decision point after that is worth 90.25% of the original and so on. Calculation of the value of the index in different scenarios is somewhat involved, meaning its use is likely to be impractical at all decision points in the situations we will come up against, however, the insight it gives us into how we approach these explore-exploit problems is more valuable than just the values it takes.

Firstly, the index rewards exploring heavily relative to the already known; for example, a choice that you have no experience of, that is no successes and no failures, is likely to have a higher index than one that you have already had a reasonable amount of success with. This seems optimistic but it can be justified by the fact that it is quite easy (and quick) to discount directions that will lead to a high failure rate, as you will get failures very early on. Once you have a few directions that are giving you success most of the time, these become the preferred choices over exploring further.

The second thing we can learn is that as we increase the value of the future, say increasing our discounting factor from 95% to 98%, exploring becomes even more strongly preferred. Again, this can be rationalised as getting rid of the bad directions early to leave more time for exploitation of the fruitful ones in the future.

While this sounds very powerful we should not ignore the way in which human beings behave when it comes to decisions about the future relative to the present; for example, it may not be appropriate to discount with a constant factor in this way. We've discussed weighing functions of utility over time in our discussion of agents, and this can be something to check against before diving in headfirst with the Gittins index.

Secondly, many of the problems we deal with on a day-to-day basis do not have constant probabilities for success and failure. This may be true for casinos and slot machines but in the real world, the chances of success are determined by the objective and the resource available to achieve it. On a similar line, an avenue that delivers high returns early on is unlikely to continue giving the same return at a constant rate; take for example drilling for oil or any other valuable substance. In the early days there were many unexplored deposits with a high probability of success when exploring, but as competition grows and the planet's surface is explored further, the easy wins start to dry up and the going gets a lot tougher.

For anyone interested in pursuing this type of framework for addressing explore-exploit problems, I would recommend running your own experiments based on your needs and understanding of the environment you are in. I would hope that you will gain a more robust approach to dealing with the unknown than you have had before.

Constraints

We've already discussed constraints in the context of the environment and the control process; to recap, these can be either natural or products of the systems we're dealing with.

In the case of the optimisation step, these constraints will relate to the adjustments we can make to the system and the resource required to do it, though are still of course constrained by the agent's resource in every respect.

For an example, let's consider that we want to replace our production line with all new machines. Ideally, we'd like this done as fast as possible so that we can start getting the benefits sooner, but that doesn't mean we'll be able to replace them all overnight. It's likely that production will have to be halted as each machine is swapped out, which will no doubt have an impact on the utility of the outcome. These elements will serve as constraints in the adjustment we plan to make.

This is something I can relate to in my own career. My team will often identify a step in one of our processes that we think is inefficient and in need of improvement, which will typically involve interaction with the software group, who are usually snowed under fixing bugs and adding features to the software that is used across the business. The improvements we request will take a finite time to code and release, provided there is an engineer available to take it on. These adjustments can end up taking days or weeks, while you still working with the old, inefficient process.

Again, we can split our system constraints into ones that are under the ownership of the agent in the control process and those that are better described as environmental. The strength of the constraint is likely to be lower if it's under the ownership of the agent, as there might be things we can move around to avoid it, but there will be little we can do about constraints imposed on us from outside.

We should be able to incorporate these constraints into our optimisation process. We've touched upon this when discussing continuous optimisation but you'll be able to find plenty of literature that describes how this can be incorporated into other techniques. It's no good pretending these constraints don't exist; evaluating the different alternatives without considering the time taken to implement them, or any other complications, will lead us in the wrong direction. This will lead to a worse outcome if these constraints affect the utility gained from the task, and again, we can consider this a holistic approach to our optimisation.

Uncertainty

When we discuss uncertainty of our optimisation process, we can include all of our modelling uncertainties that we've discussed previously. We needn't describe these again here but we will add other possibilities, as well.

Finding the global optimum for processes we can come up with is a difficult task and one that is likely to be fraught with inaccuracy. It's unlikely that we can ever be certain that the direction we are heading in is the direction of the global optimum. There are two negative outcomes that we might encounter; one, that when we make adjustments we are heading towards a local optimum that is actually taking us away from the global optimum without our knowledge, and two, we could be heading for the global optimum with the current process design but there are changes coming up in the future that we will be less able to adjust to by heading in this direction. We should be aware that optimising the process for now does not necessarily protect us against all possible scenarios in the future.

We can use our machining process example again here; imagine we have enough information to tell us that we can increase the efficiency of producing a certain part by creating specialised machines that each take care of one step in the process. Producing these machines is likely to be expensive in the short term but pay back is reasonably quick in the context of the project due to increases in efficiency. Imagine we finish installing the last of these machines when the system we are producing is made illegal in its current form. We are now faced with transforming the production process again to make the necessary changes, which is likely to set us back further in terms of resource than if we had maintained the original process from the beginning.

The uncertainty will not only appear in the process of optimisation, but also in the communication of the adjustments we plan to make. Let's say we've been looking into our production processes in a factory and we've decided that we need to reallocate some people to different areas such that they have the greatest impact. A few months in and we don't seem to be seeing the benefits we thought we would. When we investigate, we find that many of the people we've reallocated have been spending time on tasks that they used to do in their old positions, and that this has come about because they were concerned that their old tasks were not being completed to a satisfactory standard by the new recruits. This has prevented the new employees from learning these skills while taking the established ones away from their new duties. All in all, we don't have the process we thought we would at this point.

These kinds of issues can happen when we reorganise individuals; they are independently minded and may not do what we expect them to do, even if they have the best intentions. We shouldn't therefore treat the adoption of a new process as something that will happen exactly as we intended. Incidentally, we can use this as mechanism for ironing out faults in our new process, so this is not an instruction to ignore the concerns of those who have not adapted as quickly as we might like.

33

Keeping it Real

We have covered some of the key ideas that we can use in our process of optimisation and we will now move onto some of the more practical aspects. When we introduced this subject, we said that optimisation was a state of mind. How do we instill a culture of optimisation into all our activities? The following topics will help us to apply ourselves such that we make the best of all our processes.

Continuous Improvement

The optimisation of processes is of course nothing new. One of the most established methods of ensuring processes are developed routinely is known as 'Continuous Improvement'. This is a process where feedback from employees is used to refine the process for increased efficiency. This tends to favor smaller incremental improvements over large, revolutionary change. Suggestions are usually made by employees at all levels via suggestion boxes or similar, which has the positive side-effect of giving employees a feeling of increased ownership of their roles and the direction they are taking. Negative feelings about their jobs are converted into both improvements in the processes and a feeling of recognition in the workforce.

This approach was particularly successful in Japanese industry in the twentieth century, eventually allowing Japanese companies to compete with well-established American and European brands. Indeed, some of the major electronics and automotive manufacturers in the world have grown up following this philosophy.

Taking an example of an automotive production line, let's say one of the employees notices that the tools they are using to attach the door to the chassis occasionally catches on their chair when they're using it. At the end of his shift, they will go to the office and drop a suggestion for a modification to the tool (or their chair) into the box. This will be read by one of the systems engineers who will then go and make the modification at the earliest opportunity, reducing the chances of the tool catching and speeding up the production slightly. After a couple more days, the same employee might notice that every time they set the tool down, they must twist awkwardly or get off their chair. Again, a suggestion will go in the box and a modification will be made. After several iterations of this process, the employee is working much more efficiently. Now multiply this by every individual on the line and you can see how it brings about significant improvements.

Something that will often surprise people is the sheer quantity of improvements that were made in Japanese companies over the course of a year and the number of suggestions employees made that led to improvements. It is possible to find evidence of staff each making several suggestions a day, totaling hundreds or even thousands a year, per employee! Take this with the high rate of adoption of these suggestions (maybe around 80%) and it is no wonder that improvements could happen so quickly.

A further refinement on the Continuous Improvement philosophy is so called 'Marginal Gains'. This was popularised by Sir Dave Brailsford, due to its huge success when he was performance director of British Cycling, where the team went from middle of the table to dominating the sport, around the time of the Beijing and London Olympic Games. This method describes a process of optimisation where hundreds of small changes, too small to have a measurable effect on performance on their own, are made in the expectation that they will culminate into real, measurable improvements in performance. Some examples could be things as simple as making sure the team's clothes are always laundered with a familiar detergent or increased use of anti-bacterial hand gel to cut down on infection and illness. Each area is seen as an opportunity for improvement. The expectation is that if your team is class of the field in every area, you are sure to come out on top during competition.

These methods are particularly relevant when dealing with well-established processes or if competing in a competitive market where the smallest advantages can be key, as is the case in competitive cycling or in the automotive sector.

In my mind, these methods can represent a slightly reductionist view. While the changes made may lead to small improvements, we cannot rely on changes at this scale to overcome fundamental flaws in the philosophy of a process. I'm sure you'll be able to find examples of projects that appear to have all the right components but fail because the ideas at their core are flawed. Multi-million dollar movies with the top directors, actors, producers, special effects teams etc. have been known to fail on the back of poor stories, while relatively unknown productions have gone on to take the world by storm due to the strong execution of their central ideas.

This links back to our optimisation hierarchy. Making a large change at the highest level of the hierarchy is likely to have a larger effect on our efficiency than smaller changes at the bottom. We must therefore ensure this mentality exists at all levels of the agent who has ownership of the task.

The key idea for me in the success of these approaches is in instilling a culture of improvement into every employee at every level. If the levels of our Complex System are constantly churning with ideas and improvements, passing up and down the layers, it must lead to a more vibrant, creative organisation. It may well be that by encouraging ownership of tasks in every employee, the performance gained through improved motivation is as big a gain as the effect of the suggestions themselves.

Some Further Notes on Optimisation

We have spent this chapter discussing optimisation in its various forms but I'd like to take a moment to discuss what optimisation definitely isn't.

In the past, psychiatrists adhered to what is known as the 'Disease Model' of psychology. In this model, mental health issues are treated like a physical disease of the mind. It was believed that by treating the disease through either talking therapies, medication or, in some cases, physical procedures, you could remove these ailments and you would be left with a normal, happy human mind.

Needless to say, this model is beginning to run its course. It has been argued that the absence of depression is not happiness and the absence of mental health problems is not normality. These ideas are pushed by proponents of so called 'Positive Psychology'. This is a more holistic approach, where the treatment is geared towards improving the patient's life, rather than (the more reductionist) removal of the 'disease' they are thought to have.

I can try to explain this with an analogy. Imagine you have broken your leg and you are needing to spend time in a wheelchair. This has brought on some depression as you now can't do some of the things you used to really enjoy. One option would be to pour a great deal of effort into healing your broken leg as fast as possible. Maybe with all the right care, you could reduce the time it takes to recover from two months to just seven weeks. This would probably take a great deal of effort and expense.

Now let's consider another alternative. You are not depressed because of your broken leg, you are depressed because of the situation you now find yourself in. While you might be confined to a wheelchair for a while, there is nothing to say you can't find new things that you enjoy just as much. Many sports have wheelchair variants, or you could consider something different, like reading something you've always wanted to read. There will be many clubs or societies that you can join that don't need full mobility. Surely this approach will bring more utility than sitting and thinking about all the things you're no longer able to do until the break has healed? This is the essence of the positive approach to optimisation. What we have to bear in mind is that removing a problem is not necessarily the most efficient thing you could do.

This kind of thinking is something that I have come across many times in my career, and it usually means the same thing; you constrain your development by trying to tackle single problems, rather than looking at the whole system. When a racing driver complains about a certain characteristic of a car, that will often become the most crucial area to address in the eyes of the team bosses. All other developments will be suspended in the hope that something can be found to 'fix' the problem in question.

In F1 there is only one measure of success, the position you finish on race day. There is nothing to say that the best handling car will win on the day, nor the one with the fewest perceived 'problems'. It's just the car with the fastest race time. By constraining your development to one narrow area, you become blind to all the other things you can do to make the car faster. Imagine you came up with something that would improve all the other areas of the car apart from the one your driver is complaining about? If it makes your expected race finishing position better, you should take it every time and not try to worry about curing the 'diseases' you believe you suffer from.

Of course, it's not just F1 that is vulnerable to this. I can put it another way. Take a look back through time at all the most successful consumer products, how many of them are completely free from 'problems'? The Model T Ford famously only came in one colour and the driver controls were not exactly how things would end up in the future. How about a Windows PC? These have sold in their billions but I would struggle to find anyone who uses one to say it's free from any problems. How about the first iPhone? It was missing a lot of the features its competitors had at the time but that didn't stop it from taking the world by storm. These products did not succeed because they were free of problems, they succeeded because of their positives.

We can link this back to our Complex Systems analogy. The performance of an F1 car, or the desirability of a PC or phone are emergent properties of their respective systems. By reducing them down to their constituent attributes, we are ignoring the levels above. It's these that we need to get right in order to achieve success.

For our task, we have reduced all the concepts that we are dealing with down to a single metric – utility. Everything we do in our optimisation process needs to increase utility, and do it efficiently. Something that 'cures' a known problem in our process should only be considered if is more efficient that implementing all other improvements we can make. Think of this as 'positive engineering' for our tasks.

34

Learning and Optimisation Summary

Our final model element has shown us how we can take a process and use our learning and understanding to make improvements and reduce our uncertainties, and we have divided our discussions into the individual processes of learning and optimisation.

We can describe learning as updating our prior beliefs based on the evidence that we've seen. We can use the Bayesian interpretation of probability to describe how our view on the world changes with every new piece of evidence that comes in, which suggests that learning is unique to the individual, and is based on the evidence that they've seen and their prior beliefs. The strength of their prior beliefs will dictate the ease at which they can be moved in the face of contradictory evidence; too open minded and we can be talked out of our previous positions with very little effort, but too rigid and we will not change our beliefs even in the face of overwhelming evidence. Both of these characteristics will mean we believe things about the universe that aren't correct.

The scientific method is the most effective means of updating our prior beliefs and is how we should try to gain understanding. When we perform formal experiments, we can make inferences based on statistics like the 'p-value', which is the probability of seeing the results we have purely by chance as a function of the noise in the system. In reality, we're going to have to make inferences on a wide range of different measurements from the environment that don't come in the form of formal experiments, for example, from observations during our daily activities and testimony of others. We can use these, together with our Bayesian updating scheme, to learn about our tasks, but the weighting we put on different pieces of evidence from different sources will be based on empirical rules. When we're performing regression-type studies we must avoid overfitting to our data if we want to use the models for prediction. While the scientific method and experimental statistics can help to reduce the possibility of making incorrect inferences from our results, there will still be some uncertainty, while our ability to measure small changes in our experiments will be limited by the power of our experiment.

At the end of our task we must be sure to review the outcome to collate all the learning we've gained, since this is the point of the task with the lowest uncertainty, where all possible outcomes have collapsed into the one we achieved. This learning should help us to understand whether the outcome that we sought out over the course of the process was as good or bad as we thought it was going to be, which is knowledge we can use to update our Utility Function for future tasks. Documenting our task, either formally or informally, will mean more of this learning is carried forward.

We can use the knowledge we've gained to make improvements in our control process through performing some optimisations, during which we hope to improve the efficiency of the control process for achieving the maximum utility. We've discussed the use of continuous or discrete optimisation schemes and the most suitable technique will be dictated by the type of parameters we're using in the optimisation. We've also introduced the concept of steepest descent/ascent to describe the direction we should move to improve our process most efficiently.

When optimising our process, we must consider the impact on other tasks that we might be involved in at the same time. We must ensure that we are maximising our *global* utility, rather than utility that is specific to one task, and we'll also benefit from making our optimisation robust. This approach describes choosing a process where the worst possible outcome is highest of all options, protecting against worst case scenarios that we can miss if we are simply optimising 'expected' utility. This will be more useful if there is significant uncertainty in our optimisation, which can come from either the process of optimisation itself, or the application of the improvements to the process.

As well as performing optimisation using methods we understand well, we can use metrics like Gittin's Index to help us decide how to weight exploration of new ideas relative to exploitation of known ones. These metrics will usually teach us that we should weight exploration heavily since discounting inefficient directions should be a quick process. The more heavily future gains are weighted, the more exploration is preferred.

There are many optimisation strategies already in use in industry, like 'Continuous Improvement' and 'Marginal Gains'. These names describe processes where making a large number of small improvements can build into significant gains in efficiency. When we use these methods, we must make sure we're not overlooking fundamental flaws in our processes in favour of small gains. Finally, the process of optimisation is not a process of removing all 'problems' from the control process - the existence of problems is not necessarily an indicator of the suitability of the process for the task we are considering and we must ensure that improvements are holistic in the context of the outcome.

Now that we have our final model element, in our next section we can begin to consider how everything we've discussed so far can be built into the final model. Here we can start to consider tasks in their entirety and set up our new approach.

Part Five

The ACE Model

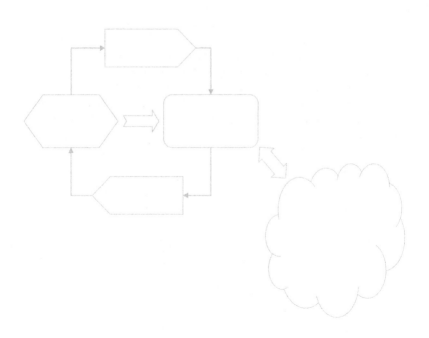

35

Building the ACE Model

Now we have all we need to start putting our model together. In the sections up to now, we've discussed all the elements we will use and touched on how they can be linked.

In our section on Agents, we've described how we can express the task owner's desires quantitatively, and how can use these desires to understand how they will behave when faced with different kinds of decisions. We also discussed the resource they have to allocate towards the task.

In the section on the Environment, we have explored the idea of Complexity and how the systems that make up the environment hierarchy will help and hinder us in achieving our desires.

The Control Process is the mechanism we will use for converting the desires of our agent into the outputs from our environment, and this will maximise their utility. We've described the kind of process this is likely to consist of, and how we might choose to act within it. Finally, our section on Learning and Optimisation taught us how we can convert this stationary process into one that's constantly updating itself with new approaches and ideas.

In each of these sections, we've shown how constraints and uncertainty will affect us when seeking out our desired outcomes; uncertainty will exist in both the elements themselves and the transfers of information between them. With all of this behind us, we can begin to appreciate the importance of each element for any task. Assembling each of these parts together will help to show us a new method with which to approach whatever we may choose to do. This is the ACE model.

Let's get to it.

The ACE Model of a Task

So here we are. In this chapter, we will put all the elements and interactions together to arrive at the completed ACE model. All the elements are of equal importance and without one, the whole model will be meaningless. The final model is described in the diagram overleaf, and the component pieces are all present with the links we have already discussed. I hope that everything feels familiar and you can begin to appreciate the meanings behind different parts of the model from our discussions of individual components.

Before we go on to discuss application of the model in its entirety, below is a quick recap of each of the elements and their interactions. Hopefully this will fill in any gaps that you may have in your memory before we go on to tackling more complex examples.

Model Components

1. Agent

- The agent is the owner of the task and contains the Utility Function and resource for the task being conducted.
- The state of the agent is evolving over time to reflect any learning or changes in resource.
- The Utility Function is used to set the target for the control process, while resource is an input to the control process and optimisation process.

274 ACE Thinking

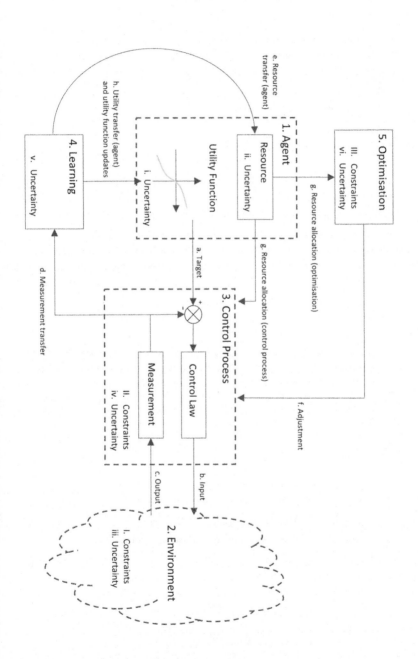

- Resource comes into the agent as an output of the learning process.
- Utility is gained and lost from inferences in the learning step, as results come in over time.
- The Utility Function can also be modified by the learning step.

2. Environment

- The environment contains all the processes that influence the task that are not under the control of the agent.
- The environment receives inputs from the control process and outputs its behavior back again.
- The environment is likely to be a complex and chaotic system.
- The internal state of the environment is evolving in time but is unknowable in the context of the task.

3. Control Process

- The control process contains the automatic response of the systems under the agent's control to achieve the desired target.
- Targets and resource are received from the agent and converted into actions on the environment.
- This also requires an estimate of the state of the environment, which is achieved through taking measurements of its outputs. These measurements are also fed into the learning process to be analysed.
- The control process can be modified by the optimisation process through any adjustments that are felt necessary.

4. Learning

- The learning step takes results from the control process, attempts to gain understanding from them, and passes this as resource to the agent.
- Utility gained during the control process is inferred and is passed to the agent also.

5. Optimisation

- The optimisation process can choose to modify the control process through knowledge gained from the learning process (via the agent's resource).
- Resource is taken from the agent in order to implement adjustments to the control process.

Component Uncertainties

When we are dealing with each of these components, we must be aware that they are affected by uncertainty. The uncertainties in each element that we need to be aware of are summarised below.

i. Utility Function uncertainty

- The Utility Function being used approximates the infinitely complex 'true' function that belongs to the agent. The true function is unknowable in the same way as the reality of the environment is unknowable.
- The act of deriving the Utility Function is analogous to performing measurements on the environment.
- We must also be aware that the Utility Function may be inaccurate if it is derived from prediction and hypothetical questioning, rather than actions being applied to the agent.

ii. Resource uncertainty

- The resource an agent has is uncertain because they cannot know the true values of the assets in their possession, nor do they completely understand the skills or knowledge of everyone working in the control process.
- The time available to perform tasks is uncertain because we cannot know how long our task will take, or whether other tasks will distract us from the one we are considering.

iii. Environment uncertainty

- Because we rely on measurements to interpret the state of the environment, the state cannot be known with certainty.
- There are also uncertainties that arise from the complex and chaotic nature of the environment. These can make it very difficult to predict the future.

iv. Control process uncertainty

- The control process may work slightly differently during each pass through even if we have similar targets and responses from the environment.
- This could be because we are forced to deviate from plans because of disturbances to the process, for example machines failing or individuals being pulled onto other tasks.
- Alternatively, mistakes can be made, or different paths through the control process might be equivalent so either can be followed without penalty.

v. Inference uncertainty

- When we make inferences on a set of data, our inferences are unlikely to describe reality perfectly.
- Uncertainty arises because there is always a chance we can misinterpret how the environment behaves. For example, conducting hundreds of experiments with a significance level of 5% will yield false positive results once in every 20 experiments.

vi. Optimisation uncertainty

- The process of optimisation is likely to require models of the process we are undertaking. These models are again unlikely to completely describe the reality of the system.
- Similarly, when performing the optimisations, we have no certainty that the step we will be asked to follow is optimal in the global sense.

- This step may be sending us towards a local minimum and away from the true global minimum.

Information Transfer Uncertainty

In a similar way to the case of the components, all the information flows in the diagram will have uncertainty associated with them. This uncertainty is a key feature of the model as it affects the way the process is played out.

a. Target uncertainty

- The uncertainty in the target arises because conveying the desires of the agent requires an additional process to simply resolving them. For example, when a manager describes a task to their employee, the requirements are likely to be incomplete or imprecise to a greater or lesser extent.
- This uncertainty will not exist if the agent happens to be the only person involved in executing the task.

b. Environment input uncertainty

- The input uncertainty arises because we cannot be certain of the action we have enacted on the environment. For example, when we throw a ball, we are not completely certain of the speed or trajectory of our arm when we release it.

c. Environment output uncertainty

- This uncertainty arises because the accuracy of our models is likely to be limited. We will not be able to predict the response from the environment precisely.
- This is similar to the uncertainty in our environment state, but it will only contain elements that we can consider as outputs.

d. Measurement uncertainty

- Our measurements are estimates of the state of the environment. Uncertainty arises because we require an additional process of measurement in the estimation. For example, when measuring the temperature of the air around us, we are reliant on the calibration of the sensor we are using.
- Any sensor is also likely to have limited accuracy/precision, meaning perfect measurements will always be impossible.

e. Resource transfer uncertainty

- When we are communicating resource changes from the learning process to the agent, there will be uncertainty if these tasks are not being performed by the agent.
- Documentation of the learning from the task may be unclear and it is unlikely that financial gains and losses can be measured precisely in real time.

f. Resource allocation uncertainty

- When allocating resource to a particular part of the control process, can we be certain that this has been achieved? For example, if asking an employee to spend more time on another project, can we guarantee that this will be carried out exactly as we expect?

g. Adjustment uncertainty

- When we order an adjustment to the control process, we cannot be certain that this has been applied in the way we would expect.
- Uncertainty arises because there is an additional process between deciding on the modifications to the control process and these being implemented. For example, if we want to make a change to the order of shift work, we must communicate this to staff. Misunderstanding can lead to different people working from what had been planned.

h. Utility transfer uncertainty

- Finally, when we communicate the results of our task, can we be sure that these are interpreted correctly into a way that gains or loses utility according to the reality of what's happened? For example, when presenting results at the end of a project, have all the relevant pieces of information been included?

Component Constraints

In each section, we have discussed when constraints will arise in our task. These can be either natural or a function of the systems involved. They can also be described as hard (inefficient to avoid) and soft (more efficient to avoid). These constraints are relevant to the following parts of the model:

   I.   Environment

- The change in state of the environment is subject to constraints that are dictated by the systems within it.
- Natural laws prevent us from doing things like creating perpetual motion and exceeding the properties of the materials available to us.
- Systems constraints will prevent us from exceeding the speed limit on the roads and catching a train at 3 am.

   II. Control Process

- The actions we can perform as part of our control process are constrained by our available resource, as well as the constraints of our environment.
- There is a subset of our system constraints that contains systems owned by the agent. These might be softer constraints than those that we will meet in the environment

III. Optimisation process

- Our optimisation process will have constraints similar to our control process, in that they include constraints of our environment and the agent. For example, we cannot replace our entire production line with new machines in a single day if the machines we require take time to design and manufacture themselves.

An Analysis of the Complete Model

With this information, we have all we need to start building the tasks we wish to complete. At first, it's unlikely that we'll be able to model the task in much detail, as we will only have simple models at our disposal. We shouldn't necessarily worry, as we often underestimate the power of simple models, and with more complex models come more parameters, greater uncertainty, and therefore more time required to maintain them. As we become more comfortable with our task we may find we have the time to start adding complexity and improving our capabilities.

For example, at the start of our modelling efforts, the agent might be pretty simple, with only a few ordered discrete preferences for utility and some financial and time resource. As we learn more, we can start to increase the complexity of this part of the model, adding continuous Utility Functions for money and indifference curves for other areas we are interested in. Now we can use these to make the decisions rather than our simple, discrete preferences. The remainder of the model could have perhaps stayed the same, but it's likely that we'll be able to make similar improvements in other areas too.

Regarding the environment, it's likely that at first we won't completely understand the systems that we're exposed to. As we discussed earlier, the world is a Complex System with many different scales we could choose to model. At the start of the process, we will be at our lowest point of understanding (provided we know how to make inferences correctly), and as we see more we can begin to learn and increase the complexity of our models.

Elements of Each Task

Now we can start to look at the processes that the model represents. These processes will play out over time, starting with the commencement of the task, and working through to the conclusion. These steps can be described as follows:

1. We first take the desires of the agent and the state of the environment and use these to determine our control actions through an initial optimisation.
2. The control acts on the environment and we measure responses, which are in turn fed back into the control process and as changes to the agent's resource (via a learning step).
3. The measurements are also fed into the learning step where inferences on their meaning take place. This is also where changes to the agent's utility are understood.
4. Learning feeds into the optimisation process, where changes to the control processes are made using the agent's resource.
5. Over time the task continues towards its eventual conclusion, where the control process ceases, and we are left with only the resource gain or loss for the entire task and any changes in the agent's utility.
6. We can apply learning from this task to other similar tasks that we may choose to embark on in the future.

In our diagram for the whole system, we can draw two distinct loops, which are shown in Figure 32. The first originates at the agent and passes to the control process as both targets and resource, then moves onto the environment as an input before going back through the control process as an output form the environment. This then travels back to the agent as resource and utility via the learning process. This loop consists of our automatic reactions to the state of the environment. If we were to draw an analogy with Daniel Kahneman's work, we could call this the 'fast' system. In this book, we will refer to this process as the 'handle-cranking' loop.

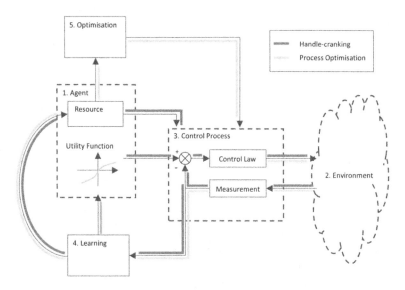

Figure 32. Loops in ACE Model

When driving a car, this loop represents the automatic responses we make to the situation we find ourselves in. In the working environment, these steps are set off as an automatic response to some input from the environment - if there is a serious fault with one of our products, an investigation is automatically initiated, or if the price of our stock falls 25% in a single day, we sell it. We should be clear that these elements are not necessarily lower skilled than those dealing with the structure of the process; indeed, in a hospital, the handle-cranking tasks are performed by highly qualified doctors and nurses, while the optimisation of the process is handled by management staff who didn't need to spend quite so long at university. These are the tasks that are closest to the outcome of the whole process, so should not necessarily be performed by unskilled individuals who can only execute instructions without understanding them.

We should think of the control process as being automatic; our reflex actions to the measurements we receive from our environment, as we've left our process of inference to the learning and optimisation step. In most circumstances, we respond to a lack of a commodity, like coffee, by ordering more of that commodity. We won't often stop to consider *why* we are ordering it every time and whether we should be looking at each of the other options available to us. Life is too short, and making these automatic decisions could well turn out to be the most rational manners in which to behave.

We can consider 'handle-cranking elements' to be part of the normal running of the task; that is, parts of the process that will happen mechanically without any additional research or input from the optimisation process.

The second loop takes resource from the agent and passes through the optimisation process, arriving in the control process in the form of adjustments. It then leaves the control process, via the environment in the form of experimental results, before passing into the learning process and finally, passing back into the agent in the form of resource. This is the 'slow' loop of the system, and it requires knowledge of the systems in the environment to make targeted adjustments to our control process. We will refer to this as our 'process optimisation' loop.

Process optimisation mostly covers indirect improvements to the outcome. This is the optimisation of the control process through experimentation and blue-sky exploration but the knowledge of which affects the shape of our Utility Function will fall under this as well. These are research tasks we can perform that reduce the uncertainty in elements of the model. A reduction in uncertainty will surely help to improve our overall performance, as now the optimal path will be clearer. We may find that we gain utility from gathering knowledge about the systems in the environment, but only because we can use this to improve the outcomes of future tasks.

Tasks that you have little experience in will typically need a good amount of process optimisation before the handle-cranking can begin. If we take the idea of buying our first pet, we should probably know more about what kind of equipment we might need, how much feeding they will require, how often they need to be let outside etc. before making the commitment. These will allow for an initial design of the process before an animal arrives on your front door needing care and attention. Again, this is not rocket science, but it is important to understand the purpose of the research we are doing and how it will help to shape the outcome of the task. Optimising our process should mean we can spend less time in the handle-cranking loop, and will, in turn, mean we can spend more time optimising.

Tasks that we are more familiar with will typically have a more established control process behind them. In this case, the process optimisation is performed in parallel to the handle-cranking. In the initial phases, improvements occur because we become more familiar with the process and it starts to become second nature; we can think of this as learning on the job. As we become more familiar, we'll find we have more time for other tasks, like doing some exploration or working to optimise the process that is already in place.

Consider your first day at work when everything probably felt quite new - you didn't know the people all that well and the layout of the building was a bit confusing but after a few weeks you began to find your feet and became more efficient in the processes you were part of. You probably reached a point where you were confident enough to start suggesting some changes to how things could be done. We can expect the same from any task we attempt for the first time.

36

Use of the Model

Now that we've shown what the model describes, we can start to understand how it can help us in improving the outcome of our task and we can describe several steps that we will go through in everything we undertake. Think of this as a template that you can apply any time you attempt something new. I hope that there are no big surprises here - you may already be familiar with a lot of the ideas we'll discuss, even if you haven't read any of the preceding sections. What I hope this approach brings is a universal template for all our tasks which is based on the model we have developed.

Firstly, we need to understand **what we want to achieve from the task**. This is our Utility Function. How do we judge whether our task is progressing successfully? Do we need to consider the opinions of multiple individuals or just one? Are we the owner of the task or do we need to understand the preferences of those that are? Maybe we need to express preferences for discrete task outcomes, or perhaps it is more appropriate to express these on a continuous scale. The answers to these questions will dictate the form of the Utility Function we will use.

This approach contrasts with some systems engineering projects, as these typically start with what the customer requires. In my opinion this is jumping the gun, since we are performing this task, whatever it may be, because we believe it will benefit ourselves. This benefit may well be indirect; for example, if we are caring for a sick family member; but we are doing these things because we gain utility from them. Any benefit to the customer is only desired because we will benefit from their satisfaction, either in the form of increased loyalty or otherwise.

Next, we need to understand **what we have at our disposal** to achieve the outcomes. This is not necessarily a fixed budget, as our Utility Function will help to define how much we are prepared to spend. We should however be aware of the constraints that are imposed on us, and the same is true for time constraints. Again, we must bear in mind that the deadlines we've been given are likely to be flexible in certain circumstances. If possible, we'd prefer the desired outcome earlier than any set deadline, which will allow us to spend more time on other tasks, gaining additional utility. Any deficiency in the skills and knowledge we require to complete the task should be considered at this point also.

We will then go through an **initial process of learning and optimisation** before we can put a control process in place. The aim is uncertainty reduction; reducing uncertainties in our model elements and information transfer will help us most at this point, before we have a control process in place. Performing this when the task is underway will take resource away from the control process, making both elements less efficient. We need to understand each element in our process and what uncertainty can trip us up along the way. At this point, any documentation from similar tasks we have performed should be consulted so that any learning is not forgotten.

When it comes to reducing our uncertainties, our model highlights the most critical types and we can use this to target how we will proceed. Some example questions that we could ask ourselves are below, organised by the element of the model they address:

Agent

- Do we understand our Utility Function as well as we could? Are there any decisions that may come up that we can use indifference curves/trade tables for? If we are executing the

task but we are not the owner, are the targets clear enough for us to plan what we are looking for? What will our measure of success be? Are there likely to be any changes to the shape of the Utility Function as the task progresses?
- Do we understand the resource available to us well enough? Is the budget we've been set really all there is, or could we get more if we can demonstrate we're doing a good job? Is there a gap in our skills that we will need to fill? How much time do we have and is this something we could ask for more of?

Environment

- What are the systems that are at play? What is the Complexity of the environment and at what scales should we be modelling?
- What are the inputs we can make to the environment from our control process and what is the effect these are likely to have? (This is our basic modelling problem)
- What is our uncertainty in the models we have?
- Is our power of prediction limited by chaotic effects? Is there anything in the future that we need to be prepared for?
- Are there other agents that are cooperating or in competition with us?
- What are the constraints that will be imposed upon us in our design of the control system and are these uncertain?

Control process

- What are the measurements we will need to take from the environment and how would we describe their uncertainty?
- What is the uncertainty in the inputs we will make to the environment?
- Do we know what experiments we will have to perform to reduce our uncertainties further?
- What information do we need to monitor and pass on for analysis?

- Could the control process itself be uncertain due to some likelihood of disruptive events or oversights?

Learning and Optimisation process

- How will we infer the success or failure of our task from the outputs of the control process?
- What are our prior beliefs about the state of the system and environment? What weighting should we give these relative to future information that we gather during the task?
- If we are doing experimentation, what significance levels and experiment powers should we be using?
- What is the uncertainty in the implementation of any adjustments we might need to make to the process? Is it just a process that needs a few people or is this a multi-disciplinary operation that will need more in place to ensure changes are made correctly?

Hopefully we can come up with answers to the above questions. The aim is simple: by reducing the uncertainty in each of our elements, we can reduce uncertainty in the outcome of the task. By tackling the areas that lead to the biggest uncertainty in the outcome, we will be able to improve it most efficiently.

If we already have some experience in the task we're performing, we should be able to put a control process in very early, particularly if it is something we have followed before or if this has been documented by a previous individual/group that has gone through the same thing. However, we should be mindful that two tasks are rarely the same and taking some time to understand any differences at the start of the task will bring greater reward than trying to make changes during the task itself. When we set off for a journey in our car and only program the Sat Nav just before we turn the ignition on, we are relying on the routes it suggests being similar to those we have experience with. How many times have we done this to find ourselves in a confusing one-way system or a toll road that we hadn't appreciated we would be directed onto? I'm sure there is a voice in the back of our heads (maybe one that sounds like a close friend or relative who is much better at planning than we are) telling us we should have anticipated something like this might happen.

Now we should be in a position to **put a control process in place** that reflects the task we are doing, which are the steps that we believe will give us the biggest pay-off over the duration of the task or when it's complete. This outcome is measured in our utility as defined by the Utility Function, which may consist of discrete or continuous elements. We would benefit from considering our optimisation loop at the same time, such that these two processes can be designed together, maybe by asking ourselves some of the following questions:

- How much resource should we allocate between handle-cranking and process optimisation tasks in order to get the best outcome? How frequently should we schedule progress reviews to take on board any learning and reflect on progress so far?
- How does the control process fit with the skills that are available to us?
- How do we ensure our control system is robust? Should we be controlling to the expected utility or a more robust level (lower 10%)?
- How can the processes we are dealing with be optimised? Are they regulator type problems that can benefit from some knowledge of the system in their design? Can we use Monte-Carlo simulations to tell us more about how to allocate resource to different areas of the task?
- What are our contingency plans if the task doesn't play out as we'd hoped?
- How will we document the process so that others can benefit from the learning we gained?

Once we have a control system in place, we can begin the **handle-cranking tasks** alongside the process optimisation. This is the part of the task we should be most familiar with. The control process is likely to be filled with jobs that we have done before, even if not necessarily in the same order. Even though the individual elements are probably familiar, we will be learning throughout this process and we should be reviewing new information regularly. Should we decide that changes need to be made, these can be fed back through the optimisation loop into the control process if suitable, or if not, documented such that we can use the knowledge when we perform this or a similar task in future.

Finally, we will reach the end of the task and **receive our outcome**. Hopefully we've been able to guide the task to a satisfactory outcome with our holistic approach, but bear in mind that even if we have lost utility we shouldn't necessarily see this as a failure, since the majority of the outcomes may have seen us lose even more. We must ensure any learning is documented such that it can be used in future tasks.

Competition and Cooperation

So far we haven't considered how competing, or indeed cooperating agents fit into our model. In reality, this is very simple; as far as we're concerned, they exist as part of our environment.

Anyone with children will know how difficult it is to get them to behave in the way that you would like them to; they are occasionally very responsive to your instructions but other times getting them to do anything is almost impossible. What seems to me to be abundantly clear is that we have no absolute authority on their behaviour, which is similar to how we might treat many other things in our environment.

If we choose to gamble in a casino there are certain actions we can employ; placing bets on roulette or playing games with elements of skill, like blackjack, but it's clear that we do not have complete control over our destinies. When dealing with children, we are typically limited to verbal instructions and (gentle!) physical persuasion. They will in turn respond however they may choose. The only difference between this and our casino example is that there is some conscious thought between our input to the environment and the output we receive. We can think of this as having two (or more) individuals separated by the environment with only their own means of communication to get the other to respond how they wish them to.

There is no mind-control here so we cannot think of a cooperating or competing agent as part of our task.

When we think of systems in the environment, we can simply include competitors and co-operators as part of these systems, and they will in turn treat us in the same way. Think of these as multiple instances of our task model, all hanging off the same environment element. Each instance of the model represents the same task belonging to a competing or cooperating agent. This is illustrated in Figure 33.

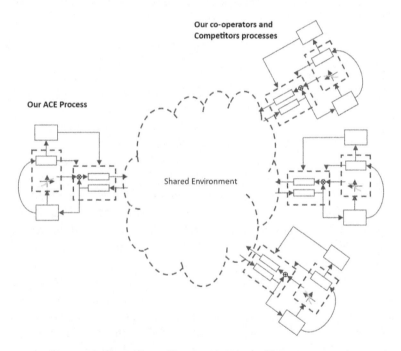

Figure 33. Illustration of how we interact with our co-operators and competitors through the environment

Let's consider another example. When choosing whether to enter a market, our environment model can include how we think the other agents in that market will respond. Chances are our competitors won't be best pleased to have increased competition and they might react strongly, with a big price drop to freeze us out, or by running a large, well targeted advertising campaign. It would be naïve of us to think that the market conditions after we enter will be the same as before we made the decision. The actions of our competitors and cooperators will need to be accounted for in our environment modelling uncertainty, as we cannot be certain how they are likely to respond. Again, we can look at Game Theory to give us some ideas, but if Behavioural Economics teaches us anything, it's that we cannot rely on agents to behave rationally. We probably don't even understand what their preferences or Utility Function looks like.

37

Examples

All of these details may sound sensible, but they are basically useless if we don't understand how they can be applied in the real world. In this section, I'll collect everything we've discussed so far into a series of simple examples that demonstrate how our new model can both describe, and help to improve on the tasks we participate in every day. The intention is to get you thinking about some of the activities in your own life, and the possible applications of this approach.

We'll choose examples from three different possible scenarios which will increase in complexity as we go. These are based on:

1. A simple task - This is owned by one person, who is also responsible for the execution the control process.
2. An industry task – This is a task that is owned by a single individual, who controls a team that are responsible for executing the control process.
3. A future task – This is a task that is owned by a group of shareholders, with a different team of people responsible for executing the control process. It will use ideas and processes that will not be possible with current practices, but this example should represent what might be achievable in the future with this model.

There are of course other permutations for how individuals could be arranged between the agent and control process, but it's hoped that the above covers all the information needed to derive the others independently. We'll avoid lengthy descriptions of complex processes or techniques in these examples for the sake of readability! There may be more detailed descriptions in other areas of the book, such as for deriving Utility Functions and indifference curves. Otherwise, you may want to seek out other pieces of literature that are dedicated to a particular subject. The intention is to showcase this approach at the highest level, rather than getting bogged down in details.

While I expect the simpler tasks we'll consider are things that you've been able to complete without any reference to ideas like Chaos and Decision Theory, I hope they can put some perspective on the subjects we've discussed so far and show how they are used as part of our model.

Simple Example

For our simple example, let's consider a task that we might take on as part of our personal lives. Your child's birthday is coming up fast and your partner has insisted that what they really want is an outdoor playset that will get them out of the house during sunny weather. Hopefully this doesn't sound like a particularly intimidating problem to begin our explorations with. Because you've spent every waking moment attending to your child's every need, there are now only a couple of weeks to buy, receive and install whatever you choose in order to be ready for the big day. We're not completely up against it, but we don't have much time to relax either.

What do we want from this task?

Let's start looking at the agent, in this case ourselves, and our preferences. What do we want from the situation? Yes, it would be great to see your child's face when they open the curtains on the morning of their birthday to reveal the world's most exciting outdoor playset, but if you don't have anything, it won't be the end of the world. Maybe we can revert to the idea of get them that games console you've had half an eye on for yourself? Or alternatively, explain that if they hadn't been quite so demanding, there would have been more time to get everything ready! We'll call the outcome of 'no outdoor playset in time for their birthday' a disutility of 100 (utility of -100). The outcome of 'outdoor playset ready for their birthday' we'll give a utility +60. This is non-symmetrical from the failure condition as we are, of course, loss averse. In this instance, we don't have any uncertainty in our targets as we are the ones owning and executing the task; we should however be aware of uncertainty in our Utility Function, as we cannot be sure exactly how we'll feel about a successful or unsuccessful outcome. Our Utility Function for money is shown below:

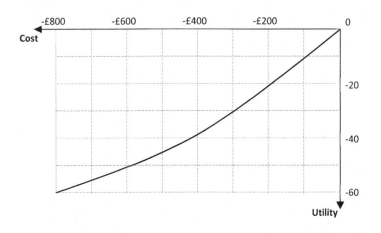

Figure 34. Utility Function for money in this example

What do we have to allocate to this task?

In terms of resource, we have around £500 to allocate towards their present. Obviously, we would prefer to spend less if we could, and we can possibly spend a little more if absolutely necessary - this is an example of uncertainty in our financial resource. In terms of time, we have all of Saturday in the first weekend and all the following weekend to buy, build and prepare it for their birthday, which is happening on the third weekend. With regard to skills, we should be capable of putting together any self-assembly system in the time suggested by the manufacturer, but we probably aren't capable of building anything from raw materials in the time available (nor would most of us want to).

Initial learning and optimisation

The environment we're working in will include the market for kids' outdoor playsets, and our house, as this is where anything we buy will be delivered, but also includes a modest selection of tools we can use to put things together with. We'll be constrained by the time it takes to get our purchase delivered, the weather on the weekends approaching (rainy days are unlikely to the best days for assembling outdoor items) and the state of the market – there has been quite a lot of sunny weather recently, so we suspect that outdoor toys will be slightly more expensive than they would be at the beginning of winter, now stocks have started to run low. The weather is a chaotic system with a prediction horizon of three to four days, so we cannot rely on the weather forecast for the following weekends. Our delivery estimate is likely to be uncertain as well.

Our control law is a function of the path we choose to take. Our actions on the environment will probably consist of placing the order for the playset and assembling it if necessary. We could buy a ready-assembled product, in which case we can get the job out of the way early and save ourselves the trouble of constructing it, or we could buy self-assembly and save a bit of money but lose time putting it together. If everything is far too expensive, we can still choose the no playset route and do our best to make amends, hopefully avoiding this being an occasion that's brought up in your child's therapy sessions when they're older. Our outputs from the environment will include things like the date the box will arrive on the back of a truck and the weather on the days we are available to assemble it.

On the Friday evening before the first weekend, we conclude that we don't know enough about what we're dealing with to be able to make good decisions. It's been a few years since we were last in the market to buy anything like this, and we'd like to learn more about what's available today and how much money we're going to have to part with. We decide to do some investigation into the possible options. This step will help us to update our prior beliefs to better match the reality of the situation we face, and we'll call it our 'uncertainty reduction' phase. For example, we know that the toys of a similar size that we already own would have been within our current budget, but these were bought several years ago and it's likely things have moved on a long way since then.

During these investigations, we narrow our options down to three avenues that dominate the others that are available to us. These are:

1. 'Sarah Space-Ranger' Basic, ready-assembled playset – The total price is around £800 and it can be delivered within a few days, leaving the second weekend completely free
2. 'Pirate Pete's' Premium self-assembly playset – The total price is £520 and should take one day to put together. The estimated delivery time is before the second weekend
3. 'Sammy Squirrel's' Budget self-assembly playset – The total price is £300 and should take two days to put together. This can be picked up straight away, with no delivery needed

Let's look at each of these options in turn, starting with the ready-assembled unit case. The price is steep but by getting out of the way now we'll gain some extra utility from having it done early (instant gratification). We might also be able to use the time we gain sitting out in the sunshine. There is also no risk of damaging it with a mistake during its assembly (uncertainty in control inputs) and a very small risk of it being delivered late (uncertainty in constraints from the environment). It meets all our requirements, but its appearance isn't great – maybe Sarah Space Ranger isn't the most inspiring character of the selection.

With the second option, we get a better-quality unit for a lower price, but we'll need to assemble it ourselves. There is now a chance we could damage something during assembly and getting it fixed will not be possible before the big day, meaning we would have paid the money (losing utility) but not have any playset to show for it. Clearly this is a very bad outcome. It should only take one day to assemble everything and we will have two available in the second weekend with which to do it. Plus, you always dreamed of being a Pirate when you were younger...

The final option is the cheapest but also requires the most assembly time. It would take two of the three days available to assemble it. Bad weather on either of those days (environment uncertainty) will mean that again, we have paid for the playset but will not have it available for the day itself, but at least in this case we would have spent less money. More steps in the assembly process will also increase the chances of making mistakes that lead to damage, and derision from your partner.

The decision is too complicated to make at face value, so we use an Even Swaps Table to look at the different options. The table we produce is similar to that in Figure 35 (note that rather than cancelling out the different dimensions, I've put utility scores in each column for conciseness).

	Sarah Space Ranger Playset	**Pirate Pete's Playset**	**Sammy Squirrel's Playset**
Cost	£800 (-60 utils)	£520 (-45 utils)	£300 (-30 utils)
Appearance	Middle (+20 utils)	Best (+30 utils)	Worst (0 utils)
Assembly complexity	None (0 utils)	One day (-10 utils)	Two days (-30 utils)
Delivery time	A few days (-5 utils)	One week (-10 utils)	Collect straight away (0 utils)
Total	-45 utils	-35 utils	-60 utils

Figure 35. Initial Even-swaps table

'Sarah Space Ranger' loses out because it's just too expensive. For the remaining options, we decide that the extra risk of needing two days to build the 'Sammy Squirrel' set means it doesn't look like a better decision than the more expensive, easier to assemble one. We therefore come down on the side of 'Pirate Pete's' boat-themed adventure playset, that you secretly wanted all along, and go ahead with the order.

Putting the control law in place

We now have our decision and the rest of our control law is laid out in front of us – after buying the playset, we must use one of the two days in the following weekend to assemble it. If it is forecast to rain on either day, we will use the other for assembly. When we're putting everything together, we should follow the instructions that are included, which will hopefully reduce the risk of oversights resulting in spending time going back to correct them.

When it comes to measurements that we might need to take when assembling, we want to make sure that each item doesn't 'wobble' when it's put in place i.e. we want all the legs to sit on the ground at the same time and not have one off the surface that will cause rocking across the other legs (anything that could lead to a trip to the A&E department is to be avoided). We design our measurement to be a test of sitting on the unit on the flattest part of the patio area, where we will be assembling everything. To reduce the uncertainty from this measurement we'll try it in a few different spots just in case the ground where we perform the first test isn't level.

While we're waiting for the box of bits to arrive, we decide to do some optimisation of the process of building it. We infer from their appearance that the three masts that rise from the centre of the boat will require the same building process, meaning it's best to assemble them at the same time before moving onto the next step. This will reduce the time needed to switch tools and the total time to assemble everything should be lower. We hope that this will also reduce the risk of damaging anything and therefore having something left out of the final assembly.

We also prepare a contingency plan for the possibility that the delivery is late, or if bad weather means there is not enough time to assemble everything. If this happens, we can get a neighbour to come around and assist. There is no guarantee that they would be available and willing (resource uncertainty) but it's something we can act on if required. We would feel bad about asking for this and it would probably cost some utility, not to mention having to return the favour in future, but probably not as much as not having the finished playset ready in time.

Handle-cranking loop

On Thursday, we receive an e-mail from 'Tim's Toys' saying our order has been delayed until late on Saturday evening, meaning we've lost a whole day of possible building time. Here is some environment constraint uncertainty come to hurt us. The forecast is for some rain on Sunday but not enough to completely write off the day, and since Sunday is within the prediction horizon for our weather system we can be reasonably confident that we will be ok. It's time to do a review of where we are, and plan how are going to act next.

We need to decide whether to call on our neighbour to come and help; calling them will save time but we would rather not bother them unless we need to. Eventually, we come round to the idea that calling the neighbour is the best thing to do, as it should make a positive outcome more likely even though the pay-off might not be as high (adding robustness). Two pairs of hands working on the assembly will speed up the task, making us more robust to oversights in the amount of time the task will take. This is important given that we only have one day to complete the assembly, with the possibility of not having everything ready for the following weekend looking ever more likely. It turns out that they are all too happy to assist, given how we were let down by the delivery.

Sunday comes around and the amount of rain is more or less in line with the forecast. The building takes slightly longer than we had planned, as one of the three masts differed slightly in its design from the other two that we had to build (this was an error in our inference). However, with the help of another person, together with our optimised plan of how to build the masts simultaneously, everything gets done before the end of the day. We confirm that none of the items will wobble on a flat surface using our special measurement technique that we designed earlier. It looks like we'll be ready with everything after all!

The following weekend arrives and when your child looks out the window on the morning of their birthday, they look absolutely delighted, before asking whether there are any squirrels inside the pirate ship...

Reviewing the outcome

Once the task is completed, we have a chance to review how it went. We create the utility table shown in Figure 36, which shows where our utility was gained and lost. It shows that overall, the outcome was successful; we believe the money we spent and the effort we went to in order to complete the task was outweighed by having everything finished and ready in time for the big day. We make a note for next time that it took a little longer than advertised to assemble and that we must be careful not to assume that because some of the items look the same, that they will require the same process to put together. We also make a note of the price we paid such that we can judge the prices of future projects against this one.

By applying our model to the process in this example, we probably haven't gained a great deal; I'm sure most of the steps we've gone through here are similar to what you would have considered if you were working through this task yourself. What I hope we've demonstrated is how all the elements of this process can be organised against the model framework that we have constructed.

Item	Utility score
Purchasing playset	-35
Getting help from neighbour	-15
Everything ready in time	+60
Total	+10

Figure 36. Utility table

We've categorised all the uncertainties we faced against where they occur in the process, and I hope that in doing this you've been able to think of some others that we've not talked about in this example. Consider how we could've lost another of our allocated working days unexpectedly; say something coming up that required our immediate attention, or we may have found ourselves in a position where the assembly required a certain tool that wasn't available to us.

We've also shown the separation between the *direct* actions towards increasing utility, such as buying and assembling the playset, and the *indirect* process optimisation steps, such as the optimisation of our assembly plan and our contingency plan to add robustness to our eventual outcome.

You may have been able to spot ways in which we could improve the task further, which is of course part of the process of learning and optimisation. We haven't considered doing much exploration beyond the initial goals - maybe it's possible to rent a much bigger and more exciting playset for such occasions at a fraction of the price?

While this is a 'toy' example of how the method can be applied, I hope you see how we can build on each of these ideas as we look into more complex problems. In the next examples, we'll consider how these elements can be used for more formal tasks, involving multiple individuals working towards a goal set by someone outside the team handling the execution.

Industry Example

Now we're going to take the ideas discussed in this section so far and apply them to the kind of task that may come up in industry. In this scenario, we work for a reasonably large sporting goods company. We run the product development department and we have been asked by the CEO to start considering the cycling market, because there has been huge growth in that sector in recent years. This isn't something we have experience in, but it is hoped that our brand is powerful enough for us to break in and gain a good return on our investment. This is of course dependent on the product that we are going to design, which is a task that the CEO is heading with the support of the shareholders. For this task, we have been given a budget, a deadline for the introduction of our first product, and the freedom to exploit the skills in our team as we see fit.

What do we want from this task?

While we will be running the team in charge of executing this task, the agent is the company CEO because they have ownership. We will therefore need to capture their preferences in our approach to bring about a satisfactory outcome. Before getting too far into the detail of the project, we try to understand the aims of the CEO more thoroughly. We begin by trying to generate a Utility Function for this project, the inputs to which are listed below:

- Financial gain from sale of the products
- Customer satisfaction
- Employee satisfaction
- Environmental impact
- Time spent on this task rather than developing other products

After appropriate questioning, we believe we come up with a rough utility curve for money plus indifference curves that cover the other aspects of the project. These are still relatively uncertain, as the trades we will be making are complex, but the function does at least give a clearer indication over what is expected from us. This has reduced our 'target uncertainty' significantly but it hasn't removed this altogether, and it is our duty to go away and evaluate possible alternatives to understand whether it is possible to get a utility 'profit' by performing this task.

What do we have to allocate to this task?

For this project, we have a fixed budget that should be enough to cover the initial costs. We should be able to go back and request additional funds when necessary, but we'll probably have to demonstrate the project is progressing well. The skills and knowledge in our group are well known to us but we'll need to be careful about how we reassign people to this new task, given that they are currently working on other projects that also have some importance.

While we have a schedule for reviewing the progress of the project, there is no fixed deadline and we're going to have to base results on the agent's preferences for instant Vs delayed gratification.

Initial learning and optimisation

After we have a clearer idea of what is expected from us, we can start to do some uncertainty reduction, which will come in the form of research of a few different kinds. We'll explain these in terms of elements of the model:

- Agent – How much time do we really have given the current workload and priorities of other projects? Is the Utility Function likely to change shape in the future?
- Environment – what is the market like now? Can we come up with a model for how many units we expect to sell based on information from our competitors? What are the expectations for quality from the consumer base? What is our prediction horizon for the cycling market?
- Control process – what are the processes we could use to complete the task? What is the impact likely to be on staff morale? How does this fit with the budget, skills and time available? Do we know how we will measure customer or employee satisfaction?

From these different pieces of research, we aim to form a series of 'prior beliefs' about the task. As the task progresses, we'll refine our direction throughout using additional pieces of research; we may also learn from areas that are outside our areas of research by chance or through observations after we have put the process in place.

Similarly, we will need to do some uncertainty reduction/optimisaiton on the information transfers that will form part of the model, as follows:

- Target uncertainty – how do we ensure this is minimised throughout the task? Is it worth reviewing the targets at regular intervals?

- Environment input uncertainty – is the input we make completely understood? Do we know how the product will be launched to the market and therefore what we need to bear in mind during development? Do our production processes cause more pollution than we think?
- Environment output uncertainty – do we understand how the environment will react to our product introduction? How will our brand value be affected?
- Measurements – do we understand the sources of uncertainty in the satisfaction measurements we have chosen to make?
- Inference – what is our threshold of significance/power for research to count as 'learning'?
- Adjustment – how will we ensure changes to our processes play out as we expect?

We use our own previous experience, plus documentation from similar studies to understand how much of our resource we should allocate to each of the above areas. We also enlist some help from the marketing and sales department to create models for the environment based on our research - these models have inputs of our product quality and price. The models return total number of sales and our impact on the satisfaction of the customer over time. We have been given some idea of the uncertainty in the models, with a suggestion that the prediction horizon is around two years, with normal market forces; however, freak unforeseen events have been known to happen that can change things dramatically.

We also create our own models of employee satisfaction based on our understanding of how our direct reports like to work, which are less formal than our environment models but are a reasonable step forward from where we were at the start of the project. We should at least understand that asking certain employees to cover weekend shifts is more likely to lead to dissatisfaction than for others. The problem of staff satisfaction will be something we must 'regulate' throughout the project.

With the models we have available, we come up with three candidates for actions we can take, given our position now, which are:

- Prestige product – possibly a new, class leading bike
- Mid-range product – a more basic bike
- Basic product – maybe a helmet or some clothing

The prestige product will give our brand a significant boost, leading to increased sales of future products. The margins are smaller and the expected profits are likely to be low or may even be negative. Satisfaction with our brand, however, is predicted to be very high if executed as expected, and our employees will enjoy the challenge of designing and developing a class-leading product. The environmental impact is likely to be low due to low sales and production volumes.

The mid-range product is likely to lead to the largest revenue, although significant process optimisation will be required to make the product in sufficient quantities to sell at a good margin; this is likely to affect employee satisfaction with the resource we have available. The environmental impact is likely to be high, with larger manufacturing processes required, with more waste generated at the end of life.

The basic product is the safe bet, with good scores for employee and customer satisfaction, with a modest return on investment and high sales volumes. The time required for the project is much lower than the other options, which would also serve as some information gathering for future products in this sector. A product that performs well is a good indication that our brand will fit in this environment, indicating we can expect strong sales in more expensive products when they are launched. The environmental impact is likely to be significant but not as high as for the mid-range option.

All these give a similar 'efficiency' in terms of utility gained versus resource spent. In terms of robustness, the basic product gives the lowest total utility return but also the highest utility at the 10% probability level (the robust level). The mid-range product gives the maximum total utility return but also has the lowest utility at the 10% level. The prestige product is thought to be the biggest risk, as the expected utility gained is lower than for the mid-range model but the utility at the 10% probability level is similar to the mid-range product. If the market were to judge that the prestige product is of poor quality relative to the customer's expectations, this will do long term damage to the brand and will possibly make future sales in this segment impossible. These options are all subject to significant modelling and inference uncertainty at this stage.

We put these options to our CEO as we think there is still some uncertainty in the targets to make the decision for ourselves. Together, it is decided that because the finances and brand power of the company aren't in a position where we can be taking big risks, we will produce the basic product first, with a view to developing new products as more information on the market is gathered - the control process for the task now becomes clearer.

Putting the control law in place

We will move a smaller proportion of our staff towards designing some protective equipment for cycling to test the water. We now need to look at how we design our control process to achieve a desirable outcome and there are several stages that we're going to have to account for. These are:

- Design of the new product
- Experimentation of materials and construction
- Testing of the fully assembled product (both internal and with the customers)
- Any design development from the results of testing
- Organisation of the manufacturing
- First production runs
- Plan for launch involving time for the marketing and sales departments to prepare

Fortunately, most of these processes we are reasonably familiar with, and there is a lot of documentation and learning that covers how we should approach each of them. We're even able to make models for how each process will progress according to some simple inputs that we have authority over. These inputs are:

- Number of people to move across to the new project and what proportion of their time they should be spending on it
- The amount of budget we allocate to each task

- The amount of time we allocate to each task
- Noise factors that we might come up against, for example failure of the first batch of physical testing, or an oversight in design of the manufacturing process that means we have to repeat the process

We can then perform some simple optimisations to give us the control process with the greatest probability of achieving the best outcome (according to our Utility Function). Once we have values of the inputs that we are happy with, we can formalise the plan, in the style of a critical path for the whole process, which includes contingencies for how we will behave in the face of problems. We have regulator-style controls in place to make sure tasks are completed on time and within budget, and these take the form of cash injections and movement of personnel into different roles if we measure progress is behind schedule. We will aim to follow this process up to a point where we consider it inappropriate, against a metric of utility of the expected outcome.

Handle-cranking loop

Over the course of the project, things seem to be progressing well. We have performed experiments to understand which of the protective materials offers the best value for money and designed the outward appearance of the product. The initial product testing gave us some strange results at first but some small modifications to the process, with increased time for experimentation, proved valuable and we now understand the initial results much better.

We then get invited to meeting with the CEO. They tell us that they are concerned about employees being poached from a new startup company that has received significant investment from a wealthy competitor. We are concerned that without improving staff satisfaction they may be tempted to leave after the project is complete, which represents a change in the shape of our Utility Function and therefore the targets that we're aiming for. We need to do more to improve the satisfaction of our staff in order to achieve the best outcome in the eyes of the agent.

Now we can revisit our optimisations of the process, with the modified Utility Function as the target. This exercise suggests changing the way we resource different areas of the project; we change the 'gains' of our staff satisfaction control strategy and add more people to the project to relieve the workload from those currently onboard. This will take a little more resource spend but given the change in weighing of staff satisfaction, it is seen to be worth the money.

We need to make the adjustments whilst ensuring the changes happen according to our prescription. We therefore increase the frequency of meetings with our direct reports such that this can be monitored, therefore reducing our adjustment uncertainty.

After a while, staff satisfaction reaches a new, higher level, which is observed over several testing periods with statistical significance. This has unforeseen benefits on the time taken to complete the project, with everything handed over to sales and marketing slightly ahead of schedule.

Reviewing the outcome

After a few sales have been made we're able to review the outcome of the project; we take the latest version of the Utility Function and estimate how we have scored relative to the original expectations. This is of course meaningless if the CEO is not in agreement, as the Utility Function is based on his ideas of what success looks like. We gather scores for the financial gain for the product line, staff satisfaction at the end of the project, customer satisfaction, environmental impact and time taken to complete and feed results into the Utility Function. The results suggest we are slightly under the expected utility predicted at the start of the project, however the change in shape part way through accounts for some of the difference. We decide the course of action we chose was the right one and that we should not necessarily make significant changes to the processes we have used.

Again, you should be able to pick out the elements of the model in this example. We've introduced some new problems compared to the first example of performing a task by ourselves, which are primarily uncertainty in the targets and in our process adjustments. While we're able to approximate the shape of the agent's Utility Function, we had to keep checking our assumptions to prevent undesirable outcomes.

Even when preparing in this way we were still met with a disturbance from the environment that ended up affecting our final utility; there was little we could have done to predict this from happening but by changing our behaviour, we were probably able to hold onto more staff than we would otherwise have been able to. This is an example of doing the best we can in all scenarios, regardless of the targets at the start of the project. We might have had a forecast at the start of the year that showed it was possible to make more money but considering that we met this disturbance, we did the right thing for the agent, even if it meant missing this target.

In this example we also created some models of the environment that we could use to help in our decisions. These models were formal, rather than the qualitative internal models we had been reliant on; doing this should mean we can predict the outcomes of paths we have little experience of. We'll develop these ideas further as we go onto discuss our final example.

Future example

In this final example, we'll look at how we might be able to design our tasks in the future, which will require acceptance of the approaches we've been discussing, plus enough time for development of the procedures for them to have advanced significantly. The example we'll discuss is likely to sound somewhat alien compared to our current working practices but perhaps, given enough time to develop new systems, we could arrive at something that looks similar. It may be that what we really end up with differs from this in many ways but what I hope to do is to stimulate some creative ways of thinking about the model we have developed and how it can be applied to the tasks we participate in daily. This may sound utopian but I'm sure the way we end up working in the future is likely to sound that way to us today, regardless of whether it's built on this method.

In this example, we're developing the racing machine of the future. We don't know whether motor racing will be able to survive in its current form; it's likely that the internal combustion engine will soon be consigned to the history books, labelled as 'something that seemed like a good idea at the time' in the same way as paddle-steamers and zeppelins have been in the past. TV viewing habits are likely to be rocked by streaming services and the rise of E-sports - we could even imagine that drone technology has advanced to such a point that these are now safely raced by competitors along aerial courses across the world. Whatever the scenario, racing has survived this long and I'm sure it's appeal will endure long into the future, so we should still be able to enjoy watching the finest drivers/pilots in the world competing against each other in custom built machines.

In our example, while we will obviously be trying to win the races, the business must at least be a sustainable enterprise; for this reason, we can imagine that our Utility Function is heavily weighted towards keeping afloat. Staying competitive is clearly a top priority, not least because it makes sustainability more likely.

Rather than estimating our Utility Function from ranking preferences or simple models and indifference curves, our Utility Function is going to be built from 'Machine Learning' of our shareholders preferences; a process which involves computer algorithms generating models for us to use rather than us having to produce these ourselves. Each shareholder is periodically queried over the direction they would like the team to take - each questionnaire is then processed, removing anomalous results, and aggregated into a single, company-wide Utility Function, which is then weighted according to everyone's shareholding. This function is built over many years and covers many different variables, including preference for profit-making, preference for instant gratification, preference for race performance and so on. To improve our algorithm further, every shareholder can evaluate the outcome of every completed task or project; this is used to reduce the uncertainty in our Utility Function, as now we have validated results for our utility metric predictions at each step.

This function is available for all to use in decision making across the team. Prior to the start of any task, we create a proposal consisting of our prior beliefs for what we expect the outcome to be, which is expressed in terms of metrics relevant to the Utility Function, such as time taken, material resource required, the necessary skills and the predicted outcome in terms of performance. All parameters are presented with uncertainties such that they can be accounted for. These estimates are based on models of the team itself and how it normally behaves, but again, these estimates are refined within a learning algorithm to correct for and regular biasing of these predictions. For example, if all the tasks we have designed take twice as long as we normally predict, this is corrected here. Similarly, if there are predictions made that stretch beyond the prediction horizon for a typical process, this is flagged such that contingency plans can be drawn and weighted according to the likelihood of them playing out.

When it comes to deciding which tasks we should be undertaking at any point in time, we can use an optimisation algorithm. Obviously the team only has a limited amount of resource, but because we have a well-defined Utility Function the optimal selection of tasks can be chosen to maximise the expected utility with the resource available. This optimisation process weights exploration projects according to a set of simulations that gives scores that look similar to those produced when using the Gittins Index. This score is based on historical exploration tasks around the area being discussed, and here we are maximising a robust measure for expected utility, such as 10th percentile expected utility.

This is some way from how we decide our tasks now. If we're lucky, there may be some cost-benefit analysis that takes place from which we can choose the ones that are likely to give the best outcome. This is likely to overlook effects like staff satisfaction and other things that we care about but that aren't convenient to express numerically, but by using utility as our measure for everything, we needn't be worried about trading commodities that are measured in different units. We can also consider using non-linear transformations, like our tendency for loss aversion.

When the relevant tasks are chosen, the project groups can set about defining the control processes further. All relevant past learning is presented automatically, such that nobody needs to go trawling through libraries to look for it, and the optimal control process is designed to maximise the Utility Function of the company in a robust way. Each step in the process has estimates for knowledge and skills required to complete it, the expected time taken and the anticipated effect on staff satisfaction. Again, all parameters have uncertainties associated with them that are based on similar tasks performed in the past, which are used to develop a raft of contingency plans that are based on how to behave in the event of the process being disturbed in some way. Any uncertainty reduction tasks are also scheduled such that they can be used to optimise the control process during regular reviews.

This kind of planning may sound a bit more familiar. Engineering companies tend to have well-established processes for design and manufacture and are used to dealing with plans of this kind. What I hope this approach adds is an assessment of utility at each step, which will make our optimisation task easier in the future. By identifying 'sinks' for utility, we might be able to target our improvements more effectively.

The modelling of the team/company itself is something I expect to become more popular in the future. Our plans rely on the elements of the task being completed at certain times; we should be able to build up a picture from history of how each element is likely to progress and therefore build an expectation of the results we expect to get.

As each task progresses, measurements from the environment and control process are monitored to ensure the task is progressing in a successful way. Any new learning is documented and stored in a central database, such that is available for all, and models are updated, with correlation orientated to the emergent properties that are being studied. Any model fudges are monitored to understand the progress of the subsystem correlation, which are then used throughout the control process to give understanding of the next set of actions.

Experiments are conducted using formal inference techniques, such as measures for statistical significance, and this is used to weight the strength of evidence from any experiment so it can be accounted for in the process of learning. Once completed, any learning is added to a central database which updates the prior beliefs of different hypotheses. The weight given to each piece of new information is based on the variance of the type of experiment conducted.

Task progress is reviewed regularly, as well as when any significant new information comes to light. This process takes information and any suggestions from all involved in the project, and if it is found that the expected outcome of the task has dropped significantly below that offered by an alternative one, the task is paused and the resource allocated to the better option.

At the end of each task, the outcome is evaluated, and all documentation collated for storage in the common database. The same process of checks against the Utility Function is performed, such that better predictions can be made in future. All individuals involved in the execution contribute their suggestions, and any resource is assessed and processed as per the company Utility Function.

Let's consider an example of a possible task that might be undertaken as part of this process - driver/pilot comments from a race event suggests our racing machine is deficient to the competition in one particular area of the circuit. Our other measurements of competitor performance confirm this deficiency and our Utility Function suggests there is a high return for improving in this area. Whatever our machine looks like it is likely to involve some aerodynamic development (as any machine working in the Earth's atmosphere travelling above a certain speed will benefit from) for which we can use any of our simulation or experimentation tools.

Proposals for solutions are plentiful, which are a good mix of suggestions from previous experience and exploration ideas; each idea is given a probability distribution for expected utility gain. This is based on expected resource requirement and pay-back, weighted according to short and longer-term gains. Let's imagine that the optimisation suggests exploration into a new idea, with a contingency of a solution based on previous experience.

The individuals are moved from their previous tasks to start on this one and resource is allocated according to the results from the optimisation routine. The control process design is conducted with help from resource management software, and uncertainties in the shape of the control process are highlighted with the critical path optimised to ensure a robust return of utility. Contingency plans are designed if the preferred outcome is not achieved at each step, some of which involve terminating the project early if the expected gains do not materialise.

One of the early steps in the process is to perform experimentation in the wind-tunnel to confirm whether the proposed changes will have the desired effect. A screening study is first conducted to reduce the number of factors to a manageable amount, before designing a more sophisticated experiment for the purposes of performing a regression. This regression will involve a version of a well-established model, which uses a double-blind testing methodology to rule out any biasing from the experimenters. Unknown factors from the environment, such as different wind speeds and directions, are included in the modelling to ensure our solution is robust to these.

Once these experiments have been conducted, inferences are scrutinised by a panel of peers who are not involved with the task. This will ensure all standard protocols have been met and the experiments can be judged as being robust for purposes of learning, and with this result our prior beliefs are updated, which in turn reduces our uncertainties.

The process is monitored regularly, with ideas for optimisation considered from all individuals involved. The continuous processes we're working with are all modelled, such that they can be optimised using Mote-Carlo simulations. This leads to revisions of the critical path at each stage as new projects appear and get assigned priorities.

Learning is documented at each step using a common document template that has been designed to make the process of compiling and reading the documentation as efficient as possible.

Once the solution is designed, it is incorporated into the design of the machine. This is then tested during one of the race events, where it is confirmed to offer an improvement in the area that we were previously felt to be deficient relative to the competition. Upon validating these results over the course of the race and subsequent races (to ensure we haven't made an incorrect inference that is specific to the first event) the utility is measured, and the Utility Function updated to reflect our new reference point.

Through all this process, we have used ideas described in the sections we've already covered but the ideas are pushed to extremes, such that *usefulness* can be maximised. I expect that what we've described will represent a significant shift in the working practices we are used to using. This process is holistic and robust and is designed to bring real improvements by optimising the processes involved; it represents a removal of arbitrary targets and a step towards doing the best job you can with the resources available.

And we definitely won't say that we're going to win the championship in the next three years.

38

We're All Snowflakes

Now we've shown that we can model tasks in this way, we can imagine the human world as being full of them. Everyone from children to the elderly, single individuals up to the scale of whole countries are performing tasks that follow this same structure. The same agent will be performing many different tasks at the same time, while they may also be working as part of a task for a different agent. It's likely that any large tasks they're participating in can be divided up into smaller ones that each adhere to this model. We can describe the world of human tasks as self-similar or 'fractal', by which we mean the behavior is independent of scale.

Nature is full of fractal structures. We can think of snowflakes, the human circulatory system, the shape of the coastline and many other naturally occurring forms as being self-similar. If you look up at the clouds, you will see the same structures exist at different scales in all of them. If you were to take a picture of one alone in the sky and show it to someone else, they wouldn't be able to tell you the size of it from the picture alone. Some of the models used in F1 rely on the asphalt structure being fractal, while the profile of the road surface the car is driving over can also be described like this. We can now describe the tasks we are involved in in the same way.

Figure 37. Snowflake and Koch Curve – both shapes exhibit self-similarity

Take for example the company that you work for; we can define the agent of this company as the shareholders, as they are the ones setting the company direction, expectations for the future and drawing dividends from the profits. You and your colleagues sit in the control process and are tasked with executing tasks according to the requirements of the shareholders (probably indirectly). If you are at management level, you'll be allocating tasks to members of your team. In this system, you are the agent and those working for you are in the control process. Your preferences and Utility Function are likely to align with the shareholders on issues that affect your employment, but at the same time, you are your own person and are likely to have your own ways of getting the best from your team. At the bottom level are the individuals tasked with executing these processes, and, again, their Utility Functions will be similar to those of their management in some respects, but they will probably have their own working practices.

We have discussed many occasions where individuals are involved in the control process but are not part of the true agents of the task. If we look only at the level of these individuals, they can each be considered to be an agent, but their preferences will be aligned to that of the level above (if they believe them to be in their best interest). A case where this isn't true might be if there was an 'operative' of a competitor working in our organisation; you can imagine that their preferences are the inverse of what the CEO of our organisation's are, and they'll be rewarded by their employer each time they cause damage to our company. When our CEO passes this operative a task, they'll probably do whatever is in their power to ensure the task does not get completed in the way that has been requested, without being caught of course.

The preferences of the agent at the top of the hierarchy typically 'cascade' through the lower levels, provided they don't meet an employee who isn't willing to play ball. When we've talked about individuals within an organisation being part of the control process, we're relying on them taking our preferences and incorporating them into how they are working; which, of course, is not guaranteed.

Each of these levels can be thought of as behaving according to this model. Goals and systems will differ at every level, but they are all working with the same processes. This is the glue that binds civilisation together, the foundation of a capitalist society. Individuals and groups of individuals will align themselves with more powerful agents, who can give them what they need to survive, but in return they will expect your hard work and cooperation.

39

ACE Model Summary

In this section, we have built the ACE model for tasks. This is a model that we can start to use to model any task we may participate in, and we have shown how the individual elements of our model can be connected together. The major elements are:

- The Agent
- The Control Process
- The Environment
- The Learning and Optimisation loop

We've summarised which parts of the model are subject to constraints and how they're going to make achieving our objectives more difficult. We've also summarised the individual uncertainties in terms of both the model elements themselves and information transfer between the elements of the model.

We've identified two loops in the model; the first is the 'fast' handle-cranking loop, consisting of the agent, control process and environment, and directly influences the outcome of the task. The second is the 'slow' process optimisation loop, which can only influence the outcome indirectly, with improvements to the control process. Balancing these is essential for maximising our expected utility.

We've introduced the process of using the model in our task. This starts with evaluating the desires and resource of the agent and then moving to an information gathering/uncertainty reduction exercise, at which point we introduced a series of questions that you can ask yourself when considering the uncertainties you'll be dealing with. It may be wise at this point to consider how you can reduce these throughout the task through experimentation and learning. Once completed, the control process can be designed, and execution can begin. Throughout the task the learning and optimisation loop can turn measurements from the environment into adjustments to our control process, and, finally, we will reach an outcome of the task, which we must understand and document.

The position of cooperating and competitive agents in our task has been highlighted, and these exist as other agents that we can interact with through the environment. With these definitions we can begin to build a picture of the world of tasks involving human beings, which is a fractal structure where this model exists at all scales.

We have used these ideas when discussing examples from the real world, and hopefully they have demonstrated how this approach can be applied. Were we to take these methods beyond what is achievable now, we should hopefully find we can reach even greater efficiency through development of these techniques.

In our final section, we'll look at what methods we can use in practice to frame our problems in this way, and spend some time thinking about why this approach may still lead to a bad outcome, but that this is not a reason for dismissing it.

Part Six

Practical Implementation

40

You Can Use This Straight Away

We now have our completed model. We understand all the individual components and we've discussed how we can relate them to the elements of the tasks we are involved in. Hopefully at this point you can begin to see the usefulness of this approach in planning and executing tasks, and you would like to understand how to make the most of it in the future. In this section, we'll describe several practical steps we can take to make sure we get the best out of everything we set out to do.

Firstly, we'll introduce some of the personal qualities you might want to promote in yourself and others when approaching your activities in this way, which will serve us well on the occasions where we haven't got time to consider building the entire model. Next, we'll introduce 'Task Sheets' and explain how these can be used to flush out information that will be critical to the outcome of the task. We can then go on to consider why we still could fail to achieve our desired outcome, even with all the right pieces in place. Lastly, we'll examine a situation for which our model will not be appropriate - a fundamental problem with human society that nobody seems to be trying to address...

Attitudes to Promote

Before I sat down to consider how the different topics that I'd been studying could connect into a single model, I'd had a reasonable amount of success in my career. I'd rarely been the most intelligent person in the room, but I'd like to think I was good at understanding what was important in the jobs I was doing and delivering on those in a good timeframe. I believe that this, combined with my desire to seek out improvements wherever possible, got me to where I was before I began writing this book.

I hope that awareness of this approach can help us all to go one better, allowing us to enhance the aspects of our personalities that will benefit the tasks we participate in. We shouldn't necessarily view this model as a purely academic exercise, but instead use these ideas to guide our approach to different tasks. Changing our working practices to reflect these should have a significant impact on the outcomes we receive, even if we don't go through any formal process of analysis using the model we have created.

Under this heading, we'll discuss some attitudes that you may want to promote in yourself and in those you are working with to get the most out of this way of thinking. Hopefully you'll find them suitable for addressing the complex ideas we've discussed, rather than being complex themselves.

The start of our task will involve setting targets for what we want to achieve. We will benefit from being **introspective** in this process, as understanding our own preferences and capabilities will help us to reduce uncertainties as we make decisions and plan how we will act in future. If we do not understand ourselves, how are we supposed to choose directions that will bring us fulfillment? We need to be **honest** about our preferences; the existence of different personal qualities is less important than understanding whether we have them or not. It shouldn't matter whether we are risk averse or risk seeking, confident or conservative, ethical or morally reprehensible, provided you understand which. The decisions we make must be right for us, even if they differ from other people.

When making decisions, we must be sure to be **holistic** in our approach. The decisions we make in our lives are likely to involve trades between different desirable commodities at some level; some directions will involve sacrificing the present for the future and others trading our resource to gain a particular experience. Holism will allow us to make these trades and avoid sacrificing too much in the pursuit of our goals.

I hope that throughout this book you have understood the benefits of a holistic approach; a reductionist viewpoint is one that overlooks the emergent properties of a system and is not something that can reliably lead us to better solutions. The systems we will deal with in our environment are more than just the sum of their parts and we can't expect to make progress by considering them as anything else.

The huge complexity of our environment should teach us to be **humble** when it comes to our own understanding, and to steer clear of making bold predictions that stretch further into the future than we can realistically see. Chaotic systems will continue to surprise us, so we shouldn't believe we have insight that exceeds the mathematics of the system.

We should be **curious** about the environment around us; the more we understand about the inner working of the systems we're dealing with, the better our planning can be. When it comes to performing experiments on the environment, we must try to be **scientific** in our approach, which is not to say we need complicated techniques and scientific apparatus; we simply need to follow the scientific process as it is the best way to increase our knowledge of the area we are studying. This process isn't the property of the science subjects exclusively, and it's something that we would all benefit from applying more often. The most useful thing we can do in practice is keep an open mind about the results we receive from our experiments. Yes, having a hypothesis is important, but we shouldn't sacrifice the truth to make us feel intelligent for predicting what was causing a certain effect.

When working through the tasks we have set ourselves, our goal is to be **efficient**. We're trying to gain the highest utility, including minimising resource outlay, which is likely to involve being

pragmatic about the approaches we should take. We are likely to be constrained by the environment and our own capabilities, and taking big risks for small gains will not bring success in the long term. We will need to be **ruthless** in dispensing with sinks for our utility, as they will continue to cause harm for as long as they are around.

Despite the need for the processes we develop to be evaluated on the results they deliver, we should try to remain **process-orientated**. We will find this approach brings more lasting success than result-orientation, where we'll struggle to differentiate luck from processes that are truly efficient.

With humility will come the ability to be critical about all the processes we create, which will mean questioning every aspect to understand whether it can be improved. If we are **relentless** in our search for improvements, we should find we can bring our processes closer to the optimum more quickly. We should aim to improve a process every time we participate in it, which will prevent stagnation and the possibility of losing sight of our purpose.

Here, we will benefit from being **exploratory** in our thinking, as the best ideas in the future will need to be discovered first. Our changes should be **positive** and not simply attempts at removing problems from the systems we are dealing with. Again, this will mean being holistic in our approach.

We can illustrate our preferences for the above traits without any reference to mathematics that we might use in modelling. These ideas may be driven by a mathematical way of thinking, but they can be used without reference to an equation or derivation. These should be applicable in far more scenarios than the mathematical processes which they describe, and I would therefore hope that those who are put off by the idea of numbers are not put off in following them!

Task Sheets

If we can adapt to reflect the advice above, it's likely that we won't need to create a complete, formal model for everything we attempt. After all, you've made it this far without it...

However, there may be times when we'd like to do something more a little more carefully; maybe there is a lot riding on a task, and we want to go through the process before diving in headfirst with no real idea of how to behave. For this, I can propose the use of a 'Task Sheet', with prompts that are designed to guide your thinking towards the important elements of your task. This will be based on the ACE model and will hopefully stimulate some creative thinking in the early stages of the task, allowing you to improve your approach and discount bad directions early on.

An example of how a Task Sheet might look is shown in Figure 38. This sheet is split into three sections:

1. Model Elements
2. Learning/Optimisation Process
3. Sources of uncertainty - information transfer

Each section is then split into the different elements covered under that heading, which are intended as prompts to get you thinking around each idea. For example, the Utility Function we use will define the target we have for the task we're participating in. When designing our control process, we need to be aware of any constraints that we are going to come up against; do we need to ship anything that will take a finite time to arrive or does our process require a particular machine that is likely to be busy with other jobs for the foreseeable future? Are we looking beyond our prediction horizon for a system, meaning we need contingencies if things go wrong? The intention is that we take some time to fill in every cell of these tables. Hopefully all the headings are familiar and it's easy to relate each to an area in the model.

ACE Task Sheet

Model Elements

Agent	Control Process	Environment
Utility Function	Process Description	Models
Resource	Measurements	Scales of Interest
Material:		
Skills:	Constraints	Prediction Horizon
Knowledge:		
Time:		
Uncertainties		

Learning/Optimisation Process

Learning	Optimisation
Prior beliefs	Methods
Experiment requirements	Constraints
Review requirements	Robustness requirements
Uncertainties	

Sources of Uncertainty – Information Transfer

Target	Input	Output	Measurement	Utility Transfer	Resource

Figure 38. ACE Task Sheet

If there are cells that we think don't apply to us, we can say so. If there are any areas we need to take some time to understand better, we can make a note that this will form part of the uncertainty reduction process that we will embark on at the start of the task. If we think one area has too much detail to document in a single sheet, we can state where it is documented.

As well as completing this sheet, we'll need to decide on how often this process should be reviewed. It's likely that we will become aware of more items that could be entered onto the Task Sheet as we go through the process and it is vital that these aren't missed. Having regular reviews in place should allow for the opportunity of adding to the sheet so that we can proceed in a way that accounts for all our learning.

Finally, we can use this sheet as part of our final review process. When the outcome of the task has become clear, did we overlook anything in our initial planning that turned out to be important? Maybe we were too worried about getting one particular area right which turned out to be reasonably straight forward. Was there anything we entered onto the project sheet that we should have considered before the task had begun? This will help to reinforce any learning from the task and ensure it's all documented in anticipation of conducting similar tasks in future.

Over time, I would expect that recognising each of the model elements will come naturally, and this aspect of the planning will become more and more efficient. I expect you will be able to adapt the approach to your own needs and avoid spending time considering different aspects only for the sake of it.

You should also start to recognise whether there are deficiencies in other aspects of your planning infrastructure. For example, is your documentation library easy enough to navigate that you can find whatever you're looking for quickly? Is it possible to track progress through a task and understand how much time and resource is being dedicated to each? Do you spend enough time reviewing task outcomes and understanding what you should try to improve in future? I expect that answering these questions and more like them will bring further refinements. This is just the start of the journey and there is still plenty to learn.

41

Failure is Not an Option

The outcomes of the examples we discussed in the previous section could, of course, have come out very differently. Taking the example describing the process of product development, let's say some of the assumptions we made at the start of the project were way too optimistic. We might be looking at a massive financial loss and a loss of utility overall. If your experiences of this kind of event are anything like mine, the recommendation to completely abandon the style of work we have used to perform the task will be made at some point.

We must appreciate that the failure of a task is not necessarily evidence that you have approached it in the wrong way. Failures will still happen even if you think you have protected against everything you can think of (incidentally trying to protect against everything you can think of may be what causes the task to fail!). For me, failure is an argument for developing the processes we have introduced even further.

In each case where we find ourselves failing, we need to revisit our models for the processes and the environment that we have used throughout. Are these leading us to think we will get better results than we do? Similarly, is there something in our information transfer that is pointing us in the wrong direction, and could our experiments be biased? Are our measurements inaccurate - are we even measuring the right things? Are we able to predict as far into the future as we think we can? Maybe it's our process of learning that's off; we need a significance level and threshold for experimental power that are suitable, and we must be sure that our potentially incorrect prior beliefs are not too strong to move towards the right answer. Hopefully the layout of the model will help to point you towards the areas you should be considering.

Any ideas of the factors contributing to the failure will of course need to be included in your documentation for future tasks to draw on. These might not be in the form of conclusions, validated by experiments, but instead hypotheses that can be tested in future tasks. We are bound to learn a lot more from failure than we do from success, so we can at least make the argument that a failure on this occasion will help us to avoid the same mistakes again in future.

After all this you may find that it was just a series of unpredictable events that derailed the task. Every time we make a decision with uncertainty during the process we risk a bad outcome, and we must remember that this kind of decision cannot be considered 'right' or 'wrong', only rational or irrational. It may be that when going back and looking through the information we had available before making a decision, we see we overlooked a certain risk that should have been considered. Alternatively, we see that with the information we had, we did the rational thing but that events after the decision conspired to make this look like a mistake. This is of course not something you can ever truly protect against but hopefully it will help to know you did the best you could.

We have not really discussed the role of luck in our discussion so far, but we definitely shouldn't underestimate its influence on the outcome of our task. We could imagine that there is a possibility of going through the task with every possible uncertainty in the process going against us; we could conduct our experiments with the greatest care and get very low p-values and still make incorrect inferences from our results. We could perform optimisations on our processes using well-established techniques and still miss the global optimum. We could accurately model our environment and use it to make predictions on how things will look in the future and still get unexpected outcomes that arise from the complex and chaotic behaviours we are dealing with.

One of my beliefs is that luck comes only behind budget when deciding the final championship position of an F1 team in a season, particularly when there has been a significant rule change. My experience is that teams broadly have the same technical ability and resource; projects will be conducted with the same level of rigor and learning will progress at a similar rate. For me, the biggest difference between teams is likely to be the starting point for their car development. They could begin with a development direction that is completely justified by previous experience and theory only to find a 'local minima' where the concept can be pushed no further. Jumping to a new point in the development process comes at significant cost, making developing away from this path inefficient, which will limit the final championship position significantly.

Does this mean that the employees of this team are less intelligent than those in a team that happened to choose a direction that lead to bigger gains for longer? I don't think so - they simply didn't have luck on their side. Those who ended up with the best car will claim that the development direction was always going to lead to better results, to which you can ask why they didn't start closer to the optimal point with their first design and go even faster.

In other scenarios, you will find you will have to make decisions with incomplete information. A good example of this is not knowing precisely what the tyre performance will be like until very late in the development process, when you are finally allowed to test the following year's tyres. You can come up with robust decisions that protect against a variety of scenarios, but given how sensitive the car's performance is to the tyres, it's easy to find yourself at the wrong end of luck.

Is there a way we can do more in this example? Maybe. Our models of Complex Systems will tell us that we need to get the parameters we are most sensitive to 'right' before moving onto parameters we are less sensitive to. We can view this like a hierarchy of scales; making the right decisions at the largest scales is likely to bring the greatest benefit and we can then move down the hierarchy, choosing optimal parameters as we go. One thing that will make this process difficult is the presence of chaos, which can influence the largest scales from the very smallest. In a highly chaotic environment, like stock markets, you are more likely to need luck on your side.

42

Shared Utilities

We've tried to make the model as general as possible throughout, but despite our best efforts we are going to come across one particular situation where its application is just not appropriate. This is a big problem, not only in the context of what we've discussed so far, but for all human society...

All our examples of tasks have a single agent. We've shown how we can take preferences of an agent and turn these into targets to be followed in the control process. The outputs of this process act on the environment and we interpret our measurements of this to understand where we have been successful.

Now imagine you work in the finance department for a health authority. One day you have two requests handed to you. The first is for a large quantity of a relatively cheap drug that is designed to prolong the life of elderly patients with a certain condition by a few months. The amount you have been asked for will probably extend the life of fifty patients by around one year. Clearly this is a good thing. The other request you have is for a very expensive treatment for a single patient who is quite young. This drug may increase the life of this patient by around fifty years. Again, clearly a good thing. Now if both treatments cost the same amount but we can only do one, which one should we choose?

We would probably look at the utility gained for each patient, their friends and family and for society as a whole by prolonging the life of each group of patients. Here we will find a significant problem. How do we compare the utility of individuals when utility is an arbitrary, unitless scale? Unfortunately, we don't have an answer – yet.

This may seem surprising. Intuitively we might consider a case where the fifty patients on the cheaper treatments were extremely old and frail, with a low quality of life. If you asked them, they may even decline the treatment. On the other hand, our young patient may be a newly qualified doctor, working on research to cure hundreds and thousands of people with a particular condition. Even now we cannot make the decision objectively, using our utility metrics.

In today's society, we may choose to make a democratic vote on the decision, which seems like a reasonable way of breaking the deadlock. But there is still a problem - we have assumed equal weightings on each individual when it comes to making the decision. It's this kind of problem that leads to conservative-leaning parties winning votes over more liberal parties by offering small tax cuts at the expense of help for those in most need. The tax cuts will benefit the greatest number of people by a small amount but the cuts to the most vulnerable will impact a smaller number of people significantly. Neither one extreme nor the other is the answer; the cost of reducing taxes for the majority while causing significant harm to a few individuals is not acceptable; however, taxing everyone's income such that these vulnerable people can have a much better quality of life than those paying for it is not appealing either. Is democracy a better system than the alternatives that we've seen throughout history? Undoubtedly yes. Is democracy the answer to all of society's problems? No.

So, what could an alternative be? We need to consider weighting the votes of individuals based on their Utility Function for the issue in question. Now we're back to the same problem of not being able to compare utility across agents or individuals.

There must be something we can do to address this. If we consider that human beings have a finite life and the utility that can be gained or lost through a particular experience must be finite, then the maximum utility an individual can gain or lose in their lives is finite also. Now we're down to making comparisons between *finite* numbers - surely this cannot be as impossible as some suggest?

What we're left with is riddled with uncertainties; we don't know how long an individual will live, nor do we know how their preferences will change throughout their lives. The chaos inherent within the system will make predictions of the course of people's lives uncertain to a massive degree, although what we also know is that for the most part, people's lives end up looking quite similar.

If somebody dies in their fifties or sixties, we will consider this unexpectedly early, while people that live into their hundreds have probably done better than most. Most people will probably have serious romantic relationships, possibly involving children, at some point. They will also probably be able to earn money to pay their own way. For the unlucky few, it won't be enough to live on, while for the lucky, it will be far in excess. Can these similarities, the emergent properties of our lives, serve as things to anchor our preferences to? Could we have a system where we rank our preferences against how we predict people's lives panning out?

Perhaps the key lies in advances in the (sometimes alarming) progress made in the field of artificial intelligence; soon we may all be interacting with 'intelligent' machines on a daily basis. Many of us will already own 'home assistants' that can answer questions that we ask with our own vocabulary. Can the same AI be trained to estimate how utility between individuals may compare?

Unfortunately, even when we have the capability to compare individuals' utility, we aren't at the end of our problems; we now need to deal with how we would like society to work for us. We might have to choose between 'utilitarian' and 'egalitarian' principles. Utilitarianism advocates working towards the total sum of individuals' changes in utility, for example, if we have to harm one individual to help hundreds more, we should do it. This leads to some somewhat difficult looking conclusions - can we really advocate causing people harm whatever the benefit? On the other hand, we have 'Egalitarianism', where we only provide one individual with an increase in utility if we can afford to do the same for everyone. We can draw parallels with our maximin theory of decision making. The egalitarian choice is one that maximises the minimum utility anyone can receive; a path which could mean turning down huge gains for the vast majority if one person would be no better off.

Maths, or more precisely Decision Theory, may be able to help us in this problem; John Harsanyi has produced a proof to show that utilitarianism is the optimal means to maximise utility across society. The proof is reasonably involved and will not be reproduced here but it is thought to be robust. Of course, as with all such proofs, the strength depends on the assumptions it is built upon and this tends to be the source of most of the challenges. I invite the interested reader to look it up and decide for themselves.

Even when we have created a scale by which we can compare every *living* individual, we will also need to weight the needs of future generations with those of the current one. Not doing so will probably lead to the current generation exhausting all the world's resources in a massive effort to increase their own utility. Creating this weighting will probably need some measure of how long we expect our society to exist for, which will mean we can assign some weighting to the very last generation of humanity. Given that this is the only intelligent society we know of in the entire universe, we're going to need a bit of luck guessing this one.

Again, we are left with significant problems, but it seems to me that these are nicer problems to have than those we experience in society today. There must be answers to how much gun ownership in America benefits those that own guns responsibly relative to the pain caused when looser regulations allow someone with bad intentions to get their hands on powerful weapons. The value of tax cuts for the poorest individuals relative to the money lost by having to increase tax for the wealthiest must also have a value in utility. Similarly, weighting the cost of regulating certain life-saving research against the religious beliefs of a group of people will have a solution as well. Now we just need to design the tools we need to answer them.

If there's one thing we could be doing as a society to benefit future generations, surely a robust method of deciding what we, as human beings, should prioritise needs to be up there? As I mentioned earlier, this isn't something I can help you with today but maybe you are someone who, one day, might?

Conclusions

43

You Can Use This Anywhere

Before you started reading this book, you had a goal in mind that led you to pick it up. Maybe you wanted to research task and project planning, and you thought this would be a good source. Maybe your interest was more casual, and you were only looking for something to pass the time, or maybe someone who you work for told you to read it against your wishes and you've been speed reading through to the end! Whatever your reasons, you had a target in mind.

As you made your way through the pages, your targets may have changed slightly. If you didn't want to read it initially, I hope you have found some information in here useful. If the opposite is true and now you just want to get to the end, you haven't got far to go!

I doubt that you've had completely free choice of where and when you have been able to read it - reading books when you're at work is probably frowned upon, unless your boss has an extremely positive attitude towards personal development. Maybe reading in the car or on the train makes you travel sick, and at home there are probably a number of other things demanding your attention. Despite these constraints, your environment has given you enough opportunities to make it this far.

When you originally laid out your process for how you were going to complete it, I doubt that you aimed to read through the entire thing in a single sitting. I expect you will have planned the best times to read, maybe a chapter at a time, in your lunch hour or just before bed? No doubt there will have been times when you expected to read the book but other things got in the way that meant it wasn't possible, or you couldn't get through as much as you thought you'd be able to. Maybe you reviewed your progress after a few chapters and decided that you were going to make changes to how often you read, based on this learning.

When you did read it, I expect a lot of the contents wasn't new to you, and it fell in line with your prior expectations of what a book like this would be about. Hopefully there have been things that were unexpected and have changed your mind about certain topics. It might be that you had expected something completely different to come out from the pages, but I hope that everything has been clear in how it has been presented. As I talked about in the introduction, the purpose of this book is to explain a model for tasks such that you can use in your work and personal lives to help improve the outcome. Maybe there were some things that you didn't quite make sense to you at first reading and you had to go back to revisit them, or seek out other sources. If you have any questions, I'd very much like to hear from you.

Now you're nearing the end of your task. You've gathered all the information from these pages and you may store them in some way, either in your memory, or with a few notes on a piece of paper, or something more formal, so that you can recall them when you need them next.

I hope that the *task* of reading through this has met the targets that you set out with at the start. If not, then I hope that during reading your targets changed to accommodate the points in the book, and you now see value in looking at problems in this way. I appreciate that your resource in terms of time and energy for reading is limited and hopefully you have seen this process as efficient in increasing your satisfaction.

While the description above is probably a bit 'corny', I hope it illustrates that we can describe every task we perform in the way modelled in the book. Maybe if you set out to read something similar in the near future, you can plan it using some of the ideas we've discussed.

When I set out to document some of these ideas, I went through a very similar process to what has just been described. While I had a target and a plan for how I was going to complete it, I have definitely learned a great deal about what is needed to finish a project on this scale.

The environment in which I started planning was full of uncertainty. We were only a few weeks into the first nationwide lockdown and the scale of the pandemic was only just becoming clear. I had been furloughed from work with no guarantee of returning. My kids were at home all day and were relying on my wife and I to fill in a huge hole that had just appeared in their educations. This was a difficult time for everyone and I'm sure it won't be forgotten by those of us who lived through it.

Fortunately, things seem to be on a path back to normality. The Formula 1 season eventually started in July of 2020 and, with a crammed schedule, we made it to the end of the year without major incident. The 2022 season looks to be much closer to normal, but it's still impossible to know when we'll be going racing without being affected in some way by this awful virus.

In many ways, this highlights the futility of setting targets into the future. I would like to see a single prediction from anywhere in the world in 2019 that was realised in 2020. This method is about doing the best you can with what you have in front of you. Yes, we should be aware of what might be possible but not to the extent that we become paralysed when things don't turn out as we planned.

For me, the pandemic was an opportunity to review what it was that my family and I really wanted, and to make changes where we couldn't justify keeping things the same. Our combined 'Utility Function' has been revised based on the world as it is today, and how it will hopefully be in the future. I know that we are not the only group of people to have gone through a similar thing, and now it's a case of turning all these desires into actions and results. I hope that the approach to problems presented in this book will improve the chances of them becoming reality.

In the introduction to this the book, we discussed an example of how tasks in our lives sometimes play out. This task involved your boss asking you to produce a report on the previous year's sales figures (if you can't remember please take a minute to go back and refresh your memory!). We highlighted some problems with the process we went through and discussed how the lessons we used to judge them were only based on incoherent ideas of how we should behave in this kind of situation. Now, we should be able to revisit this example and apply our new-found learning and I hope this demonstrates how using the model can be used to make improvements.

When your boss called you off the original task you were doing, they didn't consider how the new task should be prioritised relative to one we are already performing. When we do this, we overlook the possibility that more utility could be gained by completing the original task. Having no understanding of the tasks we are already performing is an example of inference uncertainty, and in this scenario, our boss has used incomplete information in the task optimisation process to adjust our control process.

When we were asked to 'use our judgement' when deciding the scope of the task, we were presented with a clear case of target uncertainty. How can we know what to implement in our control process without completely understanding the requirements of the task and how they are weighted against each other? This of course led to us having to repeat the task, wasting our time resource.

By making unsubstantiated claims about the causes of what are very small variations in sales across the year we've made an inference based on very little statistical significance and experiment power, which is likely to move our prior beliefs of the mechanisms in the system in the wrong direction. We can think of this as learning something that has not been demonstrated to be true and something that will impact our future tasks, as now we have introduced greater uncertainty into our environment models.

When we went on to make forecasts based on this ropey inference, there was no consideration for the prediction horizon of this problem. Is it possible to predict sales of our product this far into the future? Even if it is, is our model accurate enough to make this kind of forecast?

Finally, the 3000-word documentation we produced is incredibly inefficient. By being overly long, it will inhibit learning for those performing similar tasks in future, and what's more, the report is filed somewhere that will make knowing of its existence very difficult for anyone who wasn't involved in producing it.

This section closed with a statement describing how I think we can do better than this. I hope the previous examples demonstrate that we can.

Your New Heuristics

In our introduction, we discussed the idea of heuristics and showed that while they can often be useful rules of thumb, their use can lead to problems. They can cause our approach to be too narrow, or divide us into tribes that favour different approaches. Sometimes they can allow us to hold mutually contradictory positions, which is because they exist in isolation and do not point to a coherent approach to problems. This is where our ACE model can help. With this we have a holistic process that will enable us to maximise our expected return in any task we might participate in.

Of course, use of this model is not trivial and we will probably struggle to justify its use in every conceivable situation; however, we should still be able to use the ideas it teaches us. Even if you have no reason to ever think of a task in this way again, there are still a few key takeaways that I would like you to have, which I've listed below. If you prefer, you can call them something like '20 Lessons You Should Always Remember When Carrying Out a Task'.

1. **You are performing the task because you believe it is in your best interest to complete it**. If you cannot see the purpose of what you are doing, stop doing it if you are in control, or tell the person who is if you're not.

2. **Our preferences are complex.** If someone tells you that there is only one area to consider when performing a task, such as a financial outcome, they are attempting to

oversimplify some very complex interactions. If it is your boss that has suggested you should only care about profit, they will no doubt be very happy to donate their salary towards increasing it.

3. **Everything you do is a trade**, whether it's with money, your time or something else related to the specific task. The best we can hope for is to make efficient trades that increase our utility. This is 'Holism'.

4. **We should be able to reduce our preferences into a single, unitless metric called utility**. It should be possible to reduce any experience we can think of into this unit. Using tools like 'even swaps' and 'indifference curves' will tell us what the efficient trades between commodities or experiences are.

5. **Efficiency is our key objective when deciding how to perform a task or deciding between tasks to complete first** - we want the maximum utility gain for minimal resource expense. A project that takes half the time and money but gives us the same utility is one we should favour.

6. **We should not assume that human beings will behave perfectly rationally.** Behavioural Economics teaches us that there are many ways human beings can surprise us in the choices that they make, and we should be prepared for these.

7. **Uncertainty is crucial to your success.** At the start of a task, you probably have some idea how it will play out but I would be amazed if something that you didn't expect didn't come along. Now we understand that things we don't predict can happen (in any one of the ways we have discussed in this book) we have no excuses for not planning for it.

8. **Reality is unknowable.** Our measurements can only estimate what reality looks like and we shouldn't assume that these are all correct.

9. **Complexity will be inherent in everything you do.** When we create models, we need to consider those that do the best jobs at the scales we are interested in. The most detailed models, based on the smallest scales will be a nightmare to maintain and will probably give very strange results when it comes to emergent properties at the highest scales.

10. **We should not be put off by 'simple' models.** Even if they are not perfect in the way they behave, having something, even though we know it has limitations, will both allow us to make some predictions and will also encourage improvements by its very presence.

11. **Chaos limits our ability to see into the future.** When making plans, we should lay out our system in a way that does the best job it can in any scenario, rather than setting targets that we have no idea whether we have an ability to achieve them.

12. While many of the systems we encounter are laid out in such a way that they will lead to their inevitable destruction (think our energy uses and the Ruin of the Commons), **we should be able set up our systems in a way that leads to benefits for those around us** - it's all about setting the rules of the game to promote ethical behaviour given we know there is going to be competition. We must also look out for agents 'gaming' the rules to circumvent the purpose but adhere to the wording.

13. **Constraints will exist in a variety of ways.** Our process must be mindful of these constraints to get the best outcome.

14. **We must seek robustness in our processes wherever possible.** Given that the future is uncertain, we must choose the course of action that gives the best results when things go wrong, not the one that relies on everything going swimmingly to get the best results.

15. **We can think of learning as a series of revisions on our prior beliefs.** The strength of our prior beliefs will dictate how quickly we can expect to learn - too strong and we will never move from an incorrect belief, but too weak and we will get dragged all over the place by the slightest piece of evidence in one direction or another.

16. **Experimentation is key to our learning.** We must understand experimental planning and inference techniques so that we can inform our learning appropriately, and that concepts like statistical significance and experimental power should be at the forefront of our minds when we see some evidence from experimentation. We must stay detached from any results so as not to bias the outcome. We should be acting like scientists, not lawyers.

17. **Exploration is likely to lead to better outcomes than exploitation in the long term.** We have a huge amount of time ahead of us and can surely use some of it to consider something new. Consider how your life is going to look in ten or twenty years' time - I suspect some things will be very similar, while there will be inventions that you would not have imagined today that are part of your life every day.

18. **We must be positive in our push for improvements.** Fixating on 'diseases' will make us blind to other options for improving our system and we shouldn't assume that we haven't done the best job we could have because our system still has a small number of 'problems'.

19. **Documentation is essential to learning from previous projects** and it should be one of your first ports of call when investigating a new project and a big consideration when planning your task. Not enough time to document your findings will lead to greater uncertainty in future projects when you have already done the learning.

20. **After all our careful planning and execution based on robust learning, we may still fail.** Failure is an argument

for refining our processes more thoroughly, not abandoning a course of action altogether. Dismissing these techniques in favour of 'gut decisions' and loose processes because they are 'faster' is not a direction that can bring long term success with any regularity. We should be targeting efficiency, not speed

I hope you can see that the justifications behind these are rooted in some of the most sophisticated practices created by humans. The scientific method is surely humanity's greatest achievement, and it will continue to tell us the reality of our existence long into the future. The scientific fields have given us a huge array of tools that we can use to put this learning into action to achieve the strongest of our desires, and those of us who work with these tools every day will endeavor to improve upon them further.

These can be your new 'heuristics' to work with. This may seem hypocritical, given how we discussed this type of rule up to now, but now have a model of the system on which to hang these ideas, we can think of the above as a consistent set.

I hope they serve you well.

44

Closing

I would like to close with a few thoughts about the approach we have covered. In many ways, the approach we've derived does not give many answers, since the topic of uncertainty has been a foundation for a lot of our discussion, and it touches all areas of our lives. We must be prepared to believe that even our most basic assumptions have uncertainty associated with them, which is the scientific mindset. The laws and theorems that exist today only exist because they replaced the previous set, and it would be naïve to think that they will continue to be the best explanations for the systems in our universe forever.

This uncertainty will affect areas outside science. You will read stories of partners not just having affairs, but leading completely separate lives, with neither of their families aware of the existence of the other. Our uncertainties will hopefully not always be so dramatic, but we shouldn't be surprised when we get rain on a day forecast to be sunny.

Part of this uncertainty comes from the sheer complexity of the world - in chaotic systems, actions at the very smallest scales can explode to affect the very largest. Complex Systems consist of many layers, each with completely separate rules governing them, despite all being made of the same basic components.

What I hope to convey is a kind of humility in the face of all this complexity. We cannot be certain of how the tasks we're involved in will play out in the future but what we have described here should help us to do the best job we can, whatever the disturbances we get from our environment. We cannot treat uncertainty as a weakness, as this is a much better reflection of how the world really works than positions of certainty.

I hope you agree that what we've described in this book is not 'knowledge' but instead a way of thinking differently - I see this more as a skill than simple trivia that can be found in some non-fiction books. The purpose of this book for me is not to add this list of heuristics to the list you will already have in your head; it is to change your approach to these problems to one that is justifiable and consistent. The above lessons are not there to beat others over the head with when they do something 'incorrectly', they are there to guide your way of thinking towards a better way.

Over the course of the book we've touched on many fascinating areas of study, from Decision Theory to Design of Experiments to Complexity and Chaos; topics which are rich in ideas for improving your skills further. Thousands of books or other pieces of literature exist on each topic and I encourage you to seek out as many as you can, as this is the way you can reinforce the ideas discussed here and transfer them to your life.

You may have noticed recently that there isn't great value in 'knowledge' these days. Our smart phones can find us the answer to any 'University Challenge' question almost as quickly as the contestants can answer it. In my opinion, this is only going to get worse. Companies are no longer using huge filing cabinets to store records of previous projects and knowledge; in many cases, this information is stored in the cloud, ready to be accessed by any device you can connect to it with. Tools like internet search engines have illustrated how we can mine into this information to reach exactly what we want to know.

Don't forget that some of this technology is barely past its infancy - Google was founded in 1998, and Facebook in 2004. These are two of the largest companies in the world and their mission statements are to bring you whatever information you want as quickly as possible.

All this knowledge is a gold mine for your research, but it's only by applying it correctly that you will see a benefit, and this is where our skills need to come in. The best employees in businesses of the future are not going to be those who know the most - not knowing every detail of a particular subject area is not going to be a problem, provided we know how to get our hands on the information we require. We must then apply that knowledge to get the best out of ourselves.

I hope that you begin to feel your world view changing to be a better reflection of how it truly is. I hope that some of the topics we've covered have given you motivation to go and investigate further. I hope this then drives you to pick up more books like this and develop your skills to tackle the problems in your field of expertise.

At the start of this book is a quote from McLaren's founder, Bruce McLaren, which is something that many who know the name associate him with. This quote describes his desire to improve, and the lengths he is prepared to go to master it. This is the curse of anyone who has found their passion in life and wants to perform it to its fullest; surely the point of our careers is to collect achievements that give us satisfaction? If we refer to Prospect Theory, an extrapolation of this line of thinking is to say that the most satisfying life is not one led from a position of wealth through its entirety - it is one where we are making more gains than we are losses. Starting from nothing and gaining all that we aim to achieve over time may be the most rational approach we can take.

For me, this is the power of working in such a competitive environment. I'm often asked why I choose to work in motorsport when there may be things I could be doing that are of greater benefit to humanity. When you leave work with an Automotive Engineering degree, there is only one type of company that inspires you; only one place that can make you look forward to work on a Monday morning, and that's a motorsport team.

Formula 1 is an industry that never sits still. The desire to improve is relentless. If you can demonstrate that what you're working on can bring an improvement, it *will* become reality, be it a system for the car, a new working practice or anything else. The people working there are motivated and talented. Every piece of work you produce will be scrutinised by the best in the business to ensure high standards are maintained but also in the hope that you might have hit upon something important.

The only other industry that I can think of that challenges this for creative thinking is defense. Conflict and competition bring out individual creativity like nothing else, but wouldn't we rather live in a world that has no need for highly advanced weaponry, and conflict is resolved on a racetrack or any other sporting arena? This must be one of the healthiest forms of competition that we have as a species, and surely we can use this to drive the rest of the world forward.

I believe this is the 'point' of Formula 1. The teams that I've worked in have frequently been visited by other businesses looking to gain some insight into how a competitive team is constantly developing. Indeed, this is a key benefit for many of the sponsors whose logos can be seen covering all the cars on the grid. Formula 1 teams have inspired new practices in children's hospitals, enhancements in medical production lines and even energy saving supermarket fridges. This is a way of thinking that could benefit all the companies in the world. Yes, the politics can be frustrating and there has been some very strange behaviour from one or two of the individuals involved, but these things cannot take away from the work of the majority of the people in the industry. Their efforts are what others can look to for inspiration on how they should behave.

And occasionally there'll be some good racing to enjoy on a Sunday afternoon…

Bibliography and Further Reading

Introduction

Richard Nisbitt, *Mindware: Tools for Smart Thinking* (London: Penguin Books, 2016)

William Dunham, *The Mathematical Universe: An Alphabetical Journey Through the Great Proofs, Problems and Personalities* (New York: John Wiley & Sons, 1994)

John R. Taylor, *Introduction to Error Analysis: The Study of Uncertainties in Physical Measurements* (Sausalito, CA: University Science Books, 1997)

Simon Singh, *Fermat's Last Theorem* (United Kingdom: HarperCollins Publishers, 2012)

Agent

Daniel Kahneman, *Thinking Fast and Slow* (London: Penguin Books, 2011)

Walter Mischel, *The Marshmallow Test: Understanding Self-Control and How to Master It* (London: Corgi Books, 2015)

Dan Ariely, *Predictably Irrational: The Hidden Forces that Shape Our Decisions* (London: HarperCollins Publishers, 2009)

Williams, Thomas A., Camm, Jeffrey D., Sweeney, Dennis J., Anderson, David R., Cochran, James J. *Statistics for Business & Economics* (United States: Cengage Learning, 2016)

John S. Hammond, Ralph L Keeney, Howard Raiffa, *Smart Choices: A Practical Guide to Making Better Decisions* (Boston MA: Harvard Business School Press, 1999)

Martin Peterson, *An Introduction to Decision Theory* (Cambridge: Cambridge University Press, 2009)

Douglas R. Hofstader, *Godel, Escher, Bah: An Eternal Golden Braid* (London: Penguin Books, 1980)

Ori Brafman, Rom Brafman, *Sway: The Irresistible Pull of Irrational Behaviour* (London: Random House, 2009)

Simon Sinek, *Start With Why: How Great Leaders Inspire Everyone To Take Action* (United Kingdom: Penguin Books Limited, 2011)

Ziv Cameron, Dan Ariely, "Focussing on the Forgone: How Value Can Appear So Different to Buyers and Sellers," *Journal of Consumer Research* (2000)

https://www.ted.com/talks/dan_gilbert_the_surprising_science_of_happiness - Dan Gilbert, *The Surprising Science of Happiness*

Gregory B Northcraft, Margaret A Neale, "Experts, amateurs, and real estate: An anchoring-and-adjustment perspective on property pricing decisions," *Organizational Behavior and Human Decision Processes*, Volume 39, Issue 1 (1987): 84-97

Dan Ariely, George Loewenstein and Drazen Prelec, "Coherent Arbitrariness: Stable Demand Curves Without Stable Preferences," *Quarterly Journal of Economics* 118 (2003): 73-106

Daniel Kahneman and Amos Tversky, "Prospect Theory: An analysis of decisions under risk," *Econometrica* 47 (1979): 1-13

https://www.economicsonline.co.uk – A great resource for discussion of concepts in economics

Environment

Meadows, Donella H., Wright, Diana, *Thinking in Systems: A Primer* (United Kingdom: Chelsea Green Publishing, 2008)

Steven H. Strogatz, *Nonlinear Dynamics and Chaos: With Applications to Physics, Biology, Chemistry, and Engineering* (United States: CRC Press, 2018)

James N. Webb, *Game Theory: Decisions, Interaction and Evolution* (London: Springer, 2007)

Jamshid Gharajedaghi, *Systems Thinking: Managing Chaos and Complexity* (Burlington, MA: Elsevier, 2011)

James Gleick, *Chaos: The Amazing Science of the Unpredictable* (London: Random House, 1998)

Melanie Mitchell, *Complexity: A Guided Tour* (United Kingdom: OUP USA, 2011)

John H. Holland, *Complexity: A Very Short Introduction* (Oxford: Oxford University Press, 2014)

Richard H. Thaler, Cass R. Sustein, *Nudge: Improving Decisions About Health, Wealth and Happiness* (London: Penguin Books, 2009)

Heinz Heisler, *Vehicle and Engine Technology* (Oxford: Elsevier, 1999)

Harold Hotelling "Stability in Competition." *The Economic Journal* 39, no. 153, 41-57 (1929)

https://www.bbc.co.uk/news/explainers-51265169 - Flash crash of 6th May 2010

https://www.forbes.com/sites/johnkang/2016/12/16/samsung-will-be-apples-top-supplier-for-iphones-again-in-2017/?sh=4b8d2ce81fb0 - Apple Vs Samsung

https://www.globalcitizen.org/en/content/88-blue-planet-2-changed-david-attenborough/ - The Blue Planet effect https://ratings.food.gov.uk/ - Food Hygiene Ratings UK

Quote taken from Jack Sparrow – Pirates of the Caribbean: The Curse of the Black Pearl

Control Process

Katsuhiko Ogata, *Modern Control Engineering* (Upper Saddle River, NJ: Prentice-Hall, 2002)

Donald E. Kirk, *Optimal Control Theory: An Introduction* (Mineola, NY: Dover Publishing, 2004)

Eduardo F. Camacho, Carlos Bordons Alba, *Model Predictive Control* (London: Springer, 2013)

L. Eriksson, E. Johansson, N. Kettaneh-Wold, C.Wikstrom and S.Wold, *Design of Experiments: Principles and Application* (Stockholm: Learnways, 2000)

Herman J. C. Berendsen, *A Student's Guide to Data and Error Analysis* (Cambridge.: Cambridge University Press, 2011)

The 'Observer Effect' - The SAGE Encyclopedia of Educational Research, Measurement, and Evaluation (United States: SAGE Publications, 2018)

Li Tan, Jean Jiang, *Fundamentals of Analog and Digital Signal Processing* (United Kingdom: AuthorHouse, 2007)

Angus P. Andrews, Mohinder S. Grewal, *Kalman Filtering: Theory and Practice Using MATLAB* (Germany: Wiley, 2011)

NIST/SEMATECH e-Handbook of Statistical Methods, (http://www.itl.nist.gov/div898/handbook/)

Learning and Optimisation

Ben Goldacre, *Bad Science* (London: HarperCollins Publishers, 2009)

Ben Goldacre, *I Think You'll Find It's a Bit More Complicated Than That* (London: HarperCollins Publishers, 2015)

Brian Christian, To Griffiths, *Algorithms to Live By: The Computer Science of Human Decisions* (London: HarperCollins Publishers, 2017)

Matthew Syed, *Black Box Thinking: Marginal Gains and the Secrets of High Performance* (London: John Murray, 2015)

David H. Cropley, *Creativity in Engineering: Novel Solutions to Complex Problems* (Oxford: Elsevier, 2015)

Daniel Levitin, *The Organized Mind: Thinking Straight in the Age of Information Overload* (London: Penguin Books, 2015)

Jordan Ellenberg, *How Not to be Wrong: The Hidden Maths of Everyday Life* (London: Penguin Books, 2014)

Nate Silver, *The Signal and the Noise: The Art and Science of Predictions* (London: Penguin Books, 2012)

Pedro Domingos, *The Master Algorithm: How the Quest for the Ultimate Machine will Remake Our World* (London: Penguin Books, 2017)

Kevin Gurney, *An Introduction to Neural Networks* (London: Taylor & Francis, 2018)

James O'Brien, *How to be Right …in a world gone wrong* (London: Pengion Random House, 2018)

https://www.youtube.com/playlist?list=PLM2eE_hI4gSDnF-mEa9mrIYx7GCLQVN89 – From Data to Decisions lecture series – Chris Mack

B. F. Skinner. "'Superstition' in the pigeon," *Journal of Experimental Psychology, 38*(2) (1948), pp. 168–172

Saul A. Teukolsky, Brian P. Flannery, William H. Press, William T. Vetterling, *Numerical Recipes in C: The Art of Scientific Computing* (United Kingdom: Cambridge University Press, 1992)

Masaaki Imai, *Kaizen (Ky'zen), the Key to Japan's Competitive Success* (Singapore: Random House Business Division, 1986)

James K. Franz, Jeffrey K. Liker, *The Toyota Way to Continuous Improvement: Linking Strategy and Operational Excellence to Achieve Superior Performance* (United Kingdom: McGraw-Hill Education, 2011)

Alistair I.M. Rae, *Quantum Physics* (London: Oneworld, 2005)

Nancy Etcoff – *Happiness and its Surprises* (On the Disease Model and Positive Psychology)
https://www.ted.com/talks/nancy_etcoff_happiness_and_its_surprises

Whole model

Marcus Du Sautoy, *The Number Mysteries* (London: HarperCollins Publishers, 2011)

Practical implementation

Dan Gardner, *Risk: The Science and Politics of Fear* (London: Random House, 2008)

John C Harsanyi, "Cardinal Welfare, Individualistic Ethics, and Interpersonal Comparisons of Utility," *Journal of Political Economy* 63 (1955), pp 309-21

John C Harsanyi, "Bayesian Decision Theory, Rule Utilitarianism and Arrow's Impossibility Theorem," *Theory and Decision* 11, no. 4 (1955), pp 289-317

Conclusions

https://www.weforum.org/agenda/2020/11/formula-one-f1-innovation-ventilators-fridges/ - How F1 Technology is Changing the World

Index

A

ACE Model. 12, 22, 26, 273, 283, 321, 328, 329, 344
   Analysis 281
   Building 272
   Component uncertainties. 276
   Components 273
   Constraints 280
   Diagram 273–74
   Examples 294
   Information transfer uncertainty 278
   Loops 282
   Use 286
Agent 27–92, 12, 22, 237, 273, 287, 305
   Definition 33–34
   Rational 39, 59, 66, 87
   State 82–85
Alpine (F1 Team) 211
Anchoring 69, 70, 71, 87, 89
Apple 112
Ariely, Dan 67, 70
Artificial Intelligence 337
Aston Martin (F1 Team) 211
Attenborough, David 119
Auto-regressive models 165

B

Barichello, Rubens 95
Bayesian Updating 219, 268
Big Data 190, 207
Blue Planet (TV series) 119
BMI 155
Brabham Fan Car 255

Brailsford, Sir Dave 262
Brazilian GP 2010 94

C

Casino 42, 256, 257, 291
Chaos 123–29, 24, 26, 99, 139, 334, 337, 346, 350
   Butterfly Effect 126, 127
   Prediction Horizon .. 127, 139, 143, 297, 301, 305, 306, 313, 328, 343
Civilisation 27, 108, 213, 221, 235, 236, 320
Climate Change 11, 117
Columbus, Christopher 254
Competitors . 112, 113, 116, 118, 143, 292, 305
Complex System
   Adaptive 111, 112, 143
   Hierarchy 108–11
   Physical 111
Complexity
   Definition 106–8
Constraints ... 133, 185, 193, 204, 207, 249, 258, 301
   Control Process 192–95
   Environment 130–33
   Hard 133, 143, 203
   Natural 143, 194
   Optimisation 257–58
   System 131, 132, 133, 258, 280
Continuous Improvement 261, 262, 269
Control
   Combined 177–80

Continuous 169–77
Discrete 167–69
Control Law .. 149, 150, 151, 152, 166, 168, 171, 177, 196, 297, 300, 308
Control Process 145–208, 22, 23, 26, 85, 214, 272, 275, 277, 280, 283, 290
   Constraints 192–95
   State 195–96
   Targets 166
   Uncertainty 196–98
Control Theory ... 4, 26, 149, 163, 171, 176, 177
Cooperation .. 111, 112, 113, 114, 117, 118, 143, 288, 291, 292, 293, 320, 322
Covid-19 .... 5, 128
Creativity 127, 193, 237, 255, 263, 311, 328, 352
Critical Path Analysis ... 168, 169, 198, 309, 315, 316
Curiosity 326
Customer Satisfaction 72, 238, 248, 304, 307, 310

D

Data Mining 190
Decision Theory 58, 111, 338, 350
Decisions 58–73, 47
   maximin 59, 60, 241
   Robust 60
   Uncertainty 68–72
   Under Ignorance 60, 61
Delayed Gratification . 47, 48, 49, 57, 305
Democracy 336
Design of Experiments .. 183, 350
Disease Model (Psychology) 264
Disturbances 100, 102, 148, 170, 174, 177, 196, 203
Documentation 230–32, 268, 279, 287, 291, 306, 332, 347
Double Pendulum 124
Double-Blind Study 189, 191, 207, 316
Dumbbell Experiment Design
.... 187

E

Echo Chamber 218
Economics 34, 41, 53
Efficiency 237–39, 345
Egalitarianism 337
Elightenment (The) 221
Emergence 107, 108
Employee Satisfaction 53, 54, 304, 305, 306, 307
Endowment Effect 67, 79
Environment 92–144
   Conservation 118–19
   Constraints 130–33
   Definition 96–98
   Inputs and Outputs .. 100–105
   State 140–41
Errors
   Random 157–58, 157–58
   Systematic 156–57, 156–57
E-sports 312
Even Swaps Method 62, 299
Example
   Cycling Equipment 303–11
   Future Racing 311–17
   Outdoor Playset 295–303
Expansive Thinking (Creativity)
.... 255
Experiment 181–91
   Biasing 189, 207, 224, 313, 316
   Hypothesis 182, 190

Traps 188–91
Exploratory 255, 327
Explore-Exploit 256, 257

F

Facebook................................ 350
Failure........................ 331–34, 347
Ferrari (F1 Team) 31, 211
Filtering 161–63
Fish, Michael 21
Flash Crash, 2010 Stock market
.... 110
Flow Chart ... 151, 169, 177, 179, 197, 200, 201, 204
Ford
   Henry 254
   Model T 265
Formula 15, 7, 18, 30, 31, 78, 79, 94, 98, 106, 121, 128, 132, 135, 146, 153, 159, 160, 161, 188, 210, 211, 212, 242, 265, 318, 333, 342, 351, 352
   Qualifying 94, 95, 199, 203, 204
Fractal.............................318, 322
Frequency Domain 162, 163, 207

G

Game Theory...... 111–18, 26, 99, 134, 143, 293
Gantt Chart 168, 241
Gilbert, Dan............................. 71
Gittins Index 256, 257, 313
Global Minimum ... 251, 252, 278
Google 350
Gun control............ 219, 286, 338

H

Haas (F1 Team)..................... 211
Handle-cranking Loop. 282, 283, 284, 285, 290, 321
Hand-Steady Game................ 103
Harsanyi, John C 338
Heuristics.... 14, 15, 24, 344, 348, 350
Hilbert Transform 165
Hofstader, Douglas R............... 81
Holism 15, 36, 54, 238, 242, 246, 247, 258, 264, 269, 291, 317, 326, 327, 344
Hulkenberg, Nico 95
Humility 326
Hypothesis Testing 182, 187, 188, 207, 223, 224, 226

I

Indifference curves 44–45, 64, 87, 92
Inference
   Experimental 223–28
   Outcome 228–30
Information Gathering. 181, 307, 322
Information Transfer 83, 84, 140, 233, 278, 287, 305, 321, 328, 329, 331
Innovation 112, 237, 254, 255
Introspection.......................... 325
iPhone 265

K

Kahneman, Daniel...... 34, 64, 81, 282
Kalman Filter 165

L

Learning...... 216–23, 22, 26, 215, 275

Knowledge Vs Experience 222–23
Uncertainty 232–33
Learning and Optimisation 208–69
Leibniz, Gottfried Wilhelm 23
Line of Best Fit 186, 225
Local Minima.249, 253, 278, 333
Lorenz, Edward 125
Lottery 42, 54, 55, 59, 66, 67
Luck 327, 332

M

Machine Learning 312
Manor (F1 Team) 211
Marginal Gains 262, 269
Marshmallow Test 47
Maximin 59, 60, 241
McLaren
    Bruce 9, 351
    F1 Team 31, 211, 351
Measurement 153–60, 279
    Accuracy 156–58
    Noise 157–58
    Processing 160–65
    Relevance 154–56
Mercedes (F1 Team) 210
Model
    Correlation.17, 136, 137, 138, 143, 190, 314
    Definition 15–19
    Errors 138
    Fudges 18, 136, 143, 314
    Internal 17, 18, 57, 311
    Simulations 137
Model Predictive Control 175
Monte-Carlo Simulation 252, 290

N

Nash Equilibrium 113

Natural Selection 112, 235
Neal, Margaret A 70
Newton, Isaac 23
Non-Disclosure Agreement (NDA) 113
Normal Distribution 20
Northcraft, Gegory B. 70

O

Observer Effect 158–60
Optimal Control 175
Optimisation 234–60, 22, 26, 215, 276, 277, 281
    Analytical 247–49
    Constraints 257–58, 257–58
    Continuous 247–53
    Discrete 253–54
    Gradient-Based 249–51
    Monte-Carlo 252–53
    Search Methods 251–52
    Simple 245–47
    Uncertainty 258–60

P

PC (Windows) 265
P-Hacking 224
PID (control) . 171, 173, 175, 195
Positive Psychology 264
Positivity 327, 347
Pragmatism 35, 327
Prediction 16, 19, 98, 288
Prior Beliefs.. 184, 217, 218, 219, 220, 221, 223, 226, 228, 267, 289, 298, 305, 313, 315, 316, 332, 343, 347
Probability 59, 66, 216–17
    Bayesian interpretaton 218, 219, 226, 267
    Frequentist interpretation .. 59

Prospect Theory .. 64–68, 45, 87, 351
   Probability Weighting......... 66
   Reference point 44, 65, 66, 67, 68, 69, 84, 87, 316
P-Value . 223, 224, 226, 232, 267, 332

R

Randomisation 189
Rational 34, 58–59, 60–61
Rationality 58
Reductionism .. 15, 136, 238, 263, 264, 326
Regression ... 186, 188, 207, 225, 226, 253, 268, 316
Regulator (Control) 151, 167, 170, 177, 178, 179, 200, 201, 205, 290, 309
Relentlessness 327
Religion 14, 20, 221, 338
Resource 74–82
   Material 75–76
   Skills and Knowledge ... 76–77
   Time 77
   Uncertainty 79–82
Review 290, 330, 310–11, 315
Robustness 239–41
Ruin of the Commons ... 117, 346
Ruthlessness 327

S

Sample 183, 184, 185–86, 187
   Size 184
Samsung 112
Satisficing 72
Scientific Method 267, 348
Screening Experiment ... 188, 316
Simulation .. 16–17, 77, 120, 137, 125

Simulator, driving 8, 17, 146, 147, 185, 234
Singh Sarso, Navinder 110
Skinner, B.F. 220
Spurious Correlation 190
Standard deviation 19, 217
State 195
Statistical Significance . 183, 184, 207, 223, 224, 310, 314, 343, 347
   Significance Level 191, 277, 289, 332
Steepest Ascent/Descent 250, 268
STEM 6
Stock Market 108, 110, 128, 166, 334
Superstitions 219
System 96–99, 107, 108, 111, 131, 132, 142, 149, 263, 281
   Linear 49–50
   Non-Linear 50
   Rule Gaming 121, 346

T

Target .. 72–73, 150–51, 84, 171–74, 175, 278, 305
   Uncertainty 278, 304, 305, 343
Task Sheets 324, 328
Tasks 10–15, 10, 11, 12, 22, 241, 284, 285
   Definition 10–11
   Elements 11–12
   Multiple 241–43
   Outcome 251, 286, 330
   Prioritisation 6, 13, 81, 195–96
Time-series (data) 161
Tuning Problem 175
Tversky, Amos 64
Type 1 Error 184, 223, 224
Type 2 Error 184, 233

Tyre 50, 95, 98, 107, 111, 135, 136, 137, 156, 333
Tyrrel
   F1 Team 210
   P34 255

U

Uncertainty
   Definition........................ 19–22
USF1 (Team) 211
Utilitarianism 337, 338
Utility 36–37, 48, 275
   Definition........................ 36–37
   Diminshing marginal ... 41, 42, 57, 63, 87
Utility Function 38–57, 65, 68, 69, 71, 72, 83, 86, 88–92, 273, 276
   Appearance 40–46
   Creation 88–92
   Elements 51–55
   Incompleteness............... 55–57
   Non-Linearities 49–51
   Uncertainty..................... 68–69

V

Variance 19, 228, 315
Vote.................................... 72, 336

W

Weather ... 97, 124, 126, 297, 301
Williams (F1 Team) . 94, 95, 211, 255
   FW14B 255
Wind Tunnel 5, 78, 79, 243

Made in the USA
Las Vegas, NV
22 November 2022